HULL
HISTORICAL MOLDING CATALOG

FOX CHAPEL
PUBLISHING

ISBN 978-1-56523-385-0

Publisher's Cataloging-in-Publication Data

Hull, Brent.
    Hull historical molding catalog / Brent Hull. -- Mount Joy, PA:
Fox Chapel Publishing, c2008.

        p. ; cm.
        ISBN: 978-1-56523-385-0

        1. Moldings--Catalogs. 2. Millwork (Woodwork)--Catalogs.
    3. Architectural woodwork--Catalogs. 4. Historic buildings--Designs
and plans--Catalogs. I. Title. II. Title: Historical molding catalog.

TH2553 .H85 2008
694.6--dc22   2008

To learn more about the other great books from
Fox Chapel Publishing, or to find a retailer near you,
call toll-free 800-457-9112 or visit us at *www.FoxChapelPublishing.com*.

We are always looking for talented authors. To submit an idea, please send a
brief inquiry to acquisitions@foxchapelpublishing.com.

Printed in China

Second Printing

CONTENTS

Dear Historical Molding Enthusiast,

Thank you for purchasing the Hull Historical Molding Catalog. This catalog resulted from my research while writing a book on historic millwork. Over the years I collected over 70 original millwork catalogs from 1870-1940. Poring over these wonderful catalogs with beautiful moldings from this "Golden Age" of American millwork, I was struck by how many of these great moldings had been lost. In talking with hundreds of architects, builders, homeowners and preservationists, it became clear that there was great interest in authentic historical millwork.

THIS CATALOG IS DIFFERENT FROM ANY YOU HAVE EVER SEEN. HERE'S WHY:

• We have selected over 500 of the best and most popular designs from over 70 years of historical molding catalogs. Each molding is organized by date and style and we provide a brief description of the molding. We think you will find this catalog to be a tremendous resource.

• Our moldings are completely authentic to the originals. Most modern moldings are significantly thinner than the originals, both in overall width and in profile depth – $5/8$" today as compared to $3/4$" originally. The thinner modern moldings lack the pleasing interplay of shadow and light of historical moldings.

• The original molding catalogs offered packages of moldings for a complete room – bases, crowns, casings, etc. These packages were very popular and usually included the best-selling moldings of each era. We offer packages in a variety of styles – Colonial, Arts and Crafts, and Period Revival designs (Tudor, French, and Mediterranean).

• Very few of the historical designs are still offered by millwork manufacturers. The vast majority of the moldings from the Standard Molding Catalogs have simply gone out of production and are no longer available. The Hull Historical Molding Catalog brings them back. You will note also that many of these moldings are drawn differently—that's because we reproduced them exactly as they appeared in the original catalogs.

Those are just four reasons, but we think you will find many more. The next few pages provide more information about historical moldings. Read on, or if you are ready, just plunge into the catalog and enjoy!

Sincerely,

Brent Hull
Founder

A BRIEF HISTORY OF HISTORIC MOLDINGS

Historic moldings are derived from classical architecture, the design system that determined the shape, proportion and details on the base, column and entablature in ancient Greek and Roman buildings. From these classical details we end up with the moldings that fill our houses and buildings today. The picture below shows the elements of the classical order and how they correspond to historical moldings. The key types of moldings are:

Crown   Also called the interior cornice. In the classical order it serves as the upper part of the entablature and is the uppermost design element in the room.

Picture Mold   Corresponds to the architrave, the design element immediately above the column. It originally served as the molding from which picture frames would be hung. In design, picture molds were often used in conjunction with the crown molding to create a more detailed, fuller look at the cornice.

Chair Rail   Corresponds to the pedestal from the classical order. In historic buildings it was designed for chair backs to rest against to avoid marring the finish on the wall. The weight and size of the chair rail is established in proportion to the other design details in a room.

Base   Originally served as the support on which the pedestal of the column rested. The size and scale of the base are proportional to the rest of the wall in the room. Base moldings are often topped with a base cape and we also frequently see shoe molding at the bottom. Both of these serve to further mimic the design of a column base cap from classical architecture.

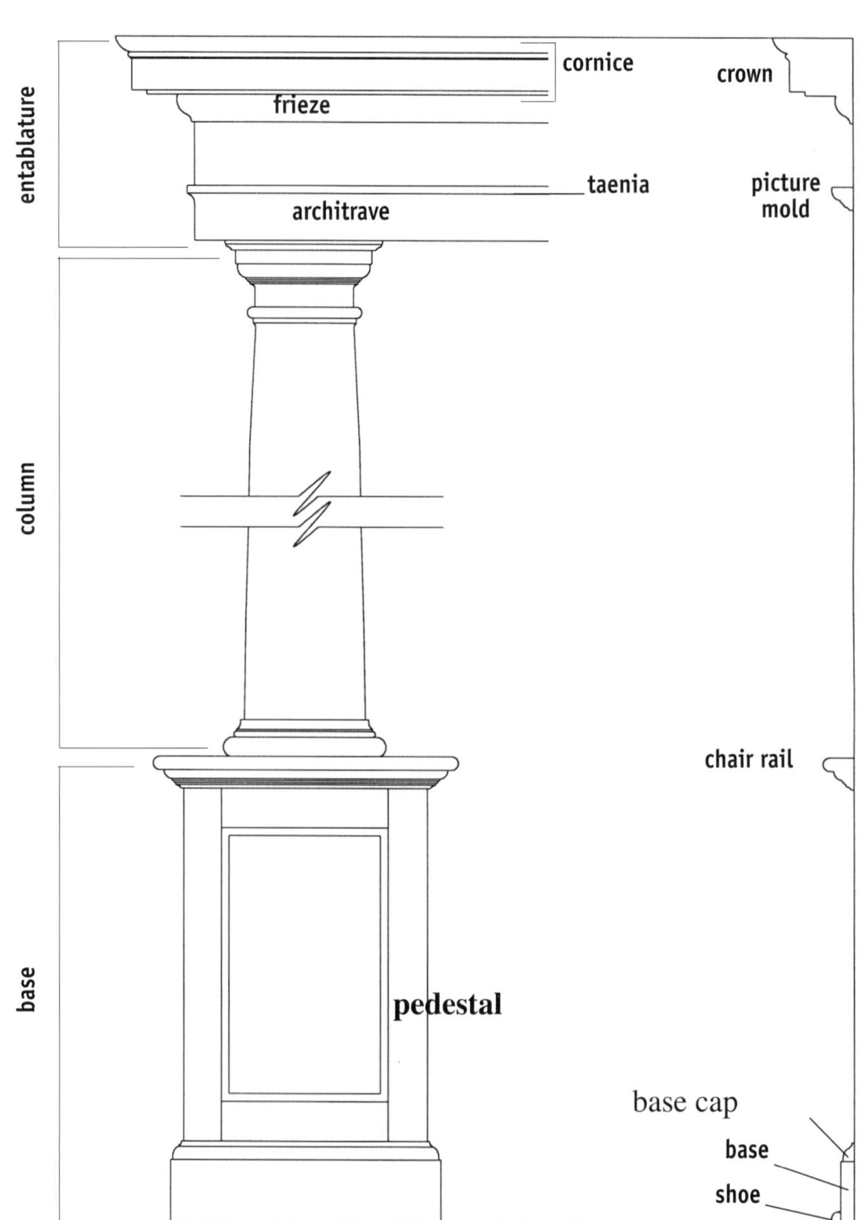

Other molding designs are not strictly tied to the classical order. Beamed ceilings, for example, are part of the Arts & Crafts tradition and are later incorporated into the Period Revival movement. The historical origin of beamed ceilings stems from early wooden construction methods and timber framing. Other miscellaneous moldings include:

Casings   The wood trim that covers the gap between a window or door and the wall.

Plinth Blocks   Square or rectangular moldings that form the bottom of a door or window casing. Usually milled with a design, but can also be plain.

Rosettes   A square molding with a circular design, usually used as a decorative feature at the corners of doors and windows.

Plate Rail   A narrow molding with a groove on the top side. It was mounted on the wall running horizontally and permitted plates or other decorative items to be displayed. Also used frequently as a molding "cap" attached to the top of wainscoting in the dining room.

Moldings in their various styles were a critical part of home design up until the 1950's. At that point, other designs such as ranch and modern moved away from moldings and they lost popularity for many years. In the 1990's, however, we saw a return to classic design and a rediscovery of many of these traditional looks. This catalog is intended not only for historical home owners, but also those building new homes who wish to capture the great warmth and charm of traditional homes.

A HISTORY OF MOLDING CATALOGS

There have not always been molding catalogs as we know them today. The first true standard molding catalog dates from 1890 when a young and growing millwork industry compiled America's first collection of moldings, which included the most popular designs of that era. This Universal Molding Catalog was the first to assign common numbers to each molding. Initially numbered 1 to 495, later the manufacturers adopted the 8,000 series numbering system. This allowed home and commercial builders and owners to purchase the same products no matter what part of the country they lived on or which mill they purchased from.

About every 10 years for the next 50 years, the millwork manufacturers introduced a revised standard molding catalog that would incorporate the latest design trends in home and commercial buildings. Each decade these standard molding catalogs would supply a snapshot of the most popular building styles. As we read catalogs from each decade we see the designs change from Victorian to Arts & Crafts, Mission, Prairie, and then to the Period Revival homes which included Colonial, Tudor, French, and Mediterranean.

As tastes changed, the new catalogs would add the latest looks and discontinue the less-popular designs. We have exhaustively researched these standard molding catalogs to determine how many years each molding was listed and also determine which style of home a molding was designed for. This research also allows us to offer many styles that were lost when they stopped being offered years ago.

HOW TO USE THIS CATALOG

The catalog is organized into sections based on type of molding. Each section has an introduction which provides a background, history and appropriate uses for each molding. The Table of Contents lists each section and where they appear. When you are ready to place an order, see our Ordering Policies on page 5.

Item Number   This is the number to enter on the quote form. Please note that some moldings are available in multiple sizes. It is critical that you enter the exact size of molding you want when you fill out the quote form.

Date   The moldings are organized by date and start with Victorian designs. The first date listed indicates when the molding initially appeared in the standard catalogs and the second date when it stopped appearing. This information will not only help you find the right molding but also to date your existing ones.

Style   Listed after the date and shows the most appropriate style home or building for this molding. As a general rule the years for each major design are: Victorian 1970-1915; Late Victorian 1900-1915; Arts & Crafts 1900-1920 (includes Craftsman, Bungalow and Prairie); Period Revival 1920-1940; and Eclectic 1930–1940. Moldings that are offered across several design trends are termed Classical. Please use the style designations as general references; some moldings exhibit characteristics of several styles.

item number ——• HHCM012

Crown Molding

date ——• 1890-1920
style ——• Classical

size ——• $^3/_4$ X $1^3/_4$

Labeled a "bed mold" in the 1890s catalog, it is part of the classical language, but this particular style had a relatively brief span of popularity. Originally available in five sizes, it stops appearing in catalogs after 1920.

Size   The first number listed is the thickness in inches, and the second number is the width. Historical moldings were often offered in a wide array of sizes and we offer you some multiple sizing options also. In these cases, the size figures are listed with thickness first and then the width options. For example, a size listing of $^3/_4$ X $1^1/_2$ or $2^1/_2$ or $3^1/_2$ or $4^1/_2$ is a $^3/_4$" thick molding available in four widths. The bigger the molding, the more expensive it will be per linear foot. This is primarily due to the greater amount of wood in larger sized moldings.

We hope this helps, and we hope you enjoy reading this catalog as much as we enjoyed putting it together!

CATALOG ORDERING POLICIES

Price quotes   You can request a price quote for any moldings in our catalog by visiting the Quotes Request Page on our web site (www.hullhistorical.com/quotes). On the Quotes Request Page simply enter the item number of the molding, the size of the molding, and the number of feet you need. Item numbers are listed in our catalog. Quote requests will be returned no later than the 2nd business day and will include all costs including shipping, setup (if applicable) and taxes.

Molding orders   Full payment must be remitted at the time of the order. No orders will be processed without payment. Payment can be made by credit card, check or money order. Standard lead times for orders are 2-4 weeks from receipt of payment until shipping. Lead times for custom orders are 4-6 weeks.

Setup charges   There are no setup charges for any moldings ordered in a quantity greater than 500 feet. Moldings which are ordered in quantities less than 500 feet are subject to a setup charge of $150 each. Multi-part moldings in order quantities of less than 500 feet are subject to the setup charge for each part.

Custom orders   The moldings in this catalog can be custom ordered in any hardwood. Prices will be quoted on request. Hull Historical can also match any molding profile if you have a molding style that is not offered in our catalog. Please call us at (800) 990-1495 for more information on our custom order program.

Custom Millwork Consulting   If you are an architect or builder looking to develop a custom millwork package for a new or historical home, we can help you select the correct moldings to meet your needs. Please call Hull Historical at (800) 990-1495 and we can provide more information on this special service.

Shipping   Standard orders (over 500 feet) are shipped LTL. Orders of less than 500 feet are usually less expensive to ship when sent via FedEx in 8-foot sections. The minimum price on LTL shipping is $65. The minimum price on FedEx shipping is $50.

Quality Assurance   Hull Historical moldings are made from clear, kiln-dried hardwood lumber. Our standard wood material is poplar, a close grained wood which is suitable for painting or staining. Our moldings are milled on computer-controlled milling machines and we are AWI quality certified. Manufacturing tolerances for thickness and width are +/- $^{1}/_{16}$". We are not responsible for changes in moldings due to environmental conditions during or after installation.

We stand behind the quality of our moldings. Please contact us for any reason if you are not satisfied with the quality of moldings you receive.

HULL
HISTORICAL
MOLDING

CROWN MOLDINGS, PICTURE MOLDS AND BEAMED CEILINGS

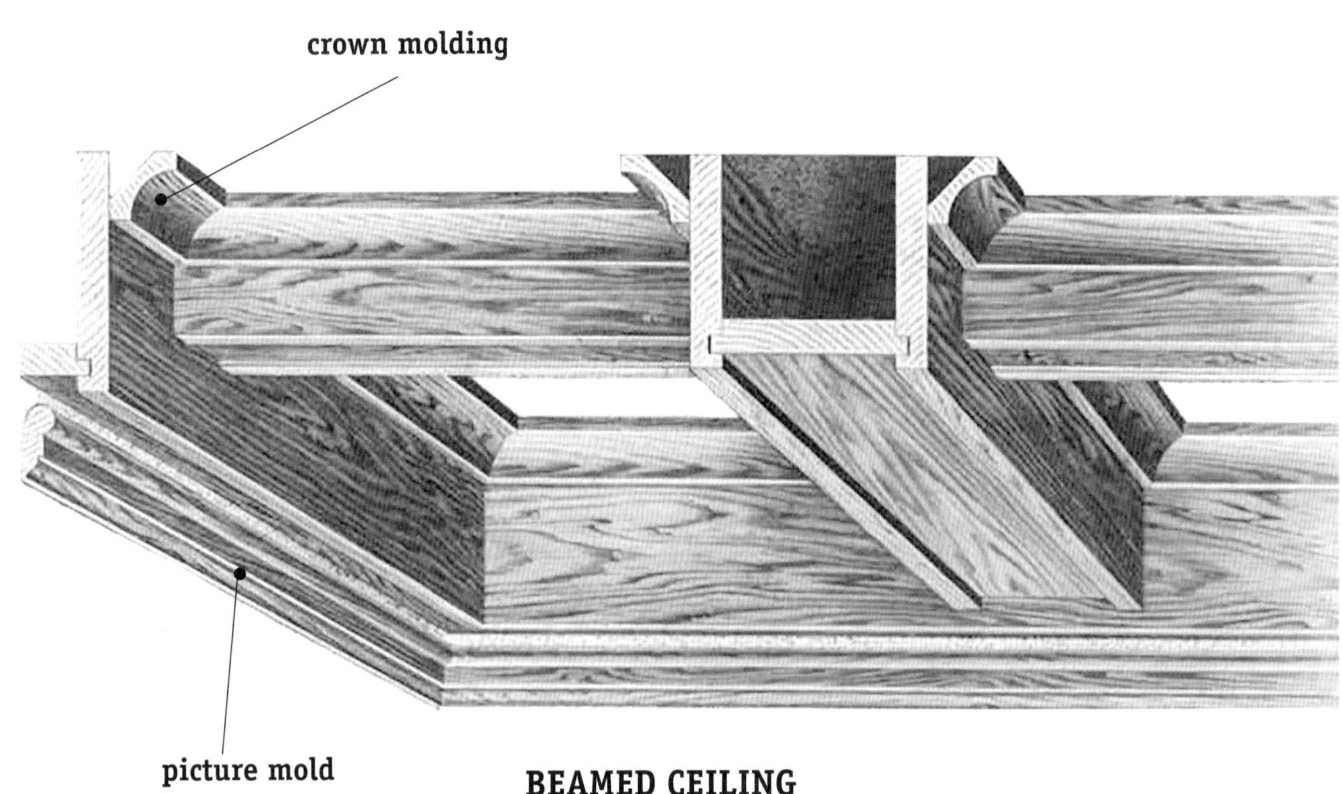

BEAMED CEILING

Crown Moldings

Originally in the 1870's, crown molds were called interior cornice moldings, which reflects their origin from classical architecture. Models like HHCM001 and HHCM004 have never gone out of style and probably never will. Others, like the Victorian crowns, may never be seen again (other than in the pages of this catalog!).

Note there is not a great diversity of crown styles; only 16 are shown here, and many of them are simply variants of each other. Yet historical home builders used these models to create great variety in looks by building them up and combining them with picture molds for more drama and impact.

Another thing you won't see elsewhere is moldings with this depth and richness of profile. Modern moldings are much thinner and have lost the beautiful interplay of shadow and light that the originals possessed —features which add great charm and warmth to a room.

HHCM001
Crown Molding
1870-1940
Classical

$^3/_4$ x $1^3/_4$ or $2^1/_4$ or $2^3/_4$ or $3^1/_2$ or $4^1/_2$ or $5^1/_2$

This was the first set of crowns, and the most popular. It was available in six sizes up to $5^1/_2$ inches—the thickness of which created nice shadows at the cornice.

HHCM002
Crown Molding
1870-1940
Classical

$^3/_4$ x $1^3/_4$ or $2^1/_4$ or $2^3/_4$ or $3^1/_2$ or $4^1/_2$ or $5^1/_2$

This style of molding is still seen today with its historical numbers 8010 and 8012. Our historically accurate thickness gives it better lines and shadows on the wall.

HHCM003
Crown Molding
1870-1940
Classical

$^3/_4$ x 1 or $1^1/_2$ or $2^1/_4$ or $2^3/_4$ or $3^1/_2$

Also called a "bed mold," this is part of the entablature in the classical series. This classical mold has many uses and can be found not only as a crown on the interior, but on the exterior. I have even seen it support a hefty cap over wainscoting.

HHCM004
Crown Molding
1870-1940
Classical

$^3/_4$ x $1^1/_2$ or $2^1/_4$ or $2^3/_4$ or $3^1/_2$ or $4^1/_2$ or $5^1/_2$

Labeled a "sprung cove" in the original 8000 series, this molding and variations of it finds popularity in the Period Revival homes of the 20's and 30's. This molding is available in six sizes for varying applications.

HHCM005
Crown Molding
1890-1914
Victorian

$^3/_4$ x $3^1/_2$

Often paired with panel molds, this Victorian crown makes it into the 1900's, but because of its Victorian styling gets cut out of the 1914 standard catalog.

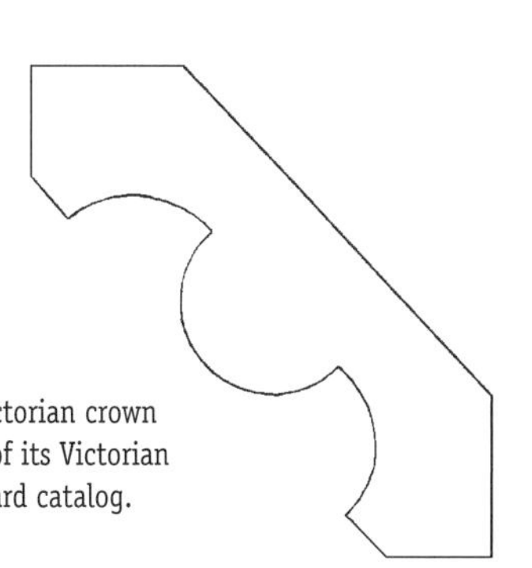

HHCM006
Crown Molding
1890's
Victorian

$1^3/_8$ x 5

This, as well as many of the Victorian crowns, can only be found in the 1890's Universal catalog. This molding has some classical stylings that could still be popular today. Note the size of the mold—quite large.

HHCM007
Crown Molding
1890's
Victorian

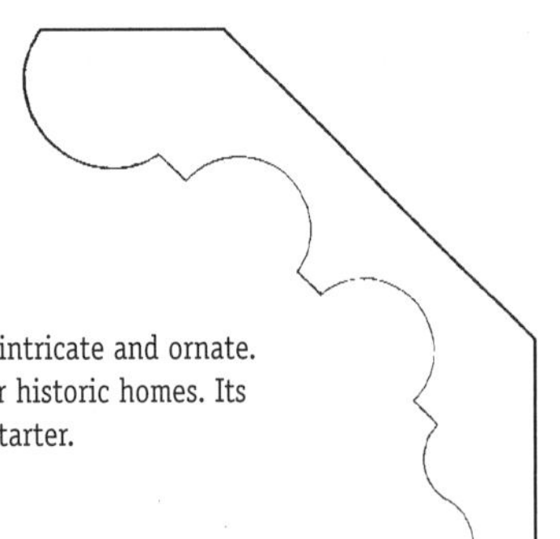

$^7/_8$ x 3

This is a typical Victorian molding, intricate and ornate.
It would be a great detail for new or historic homes. Its
unique look is a real conversation starter.

HHCM008
Crown Molding
1890's
Victorian

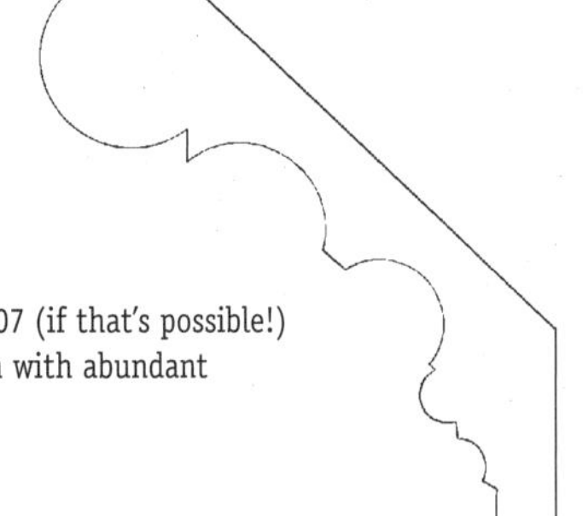

$^7/_8$ x 3

The more detailed version of HHCM007 (if that's possible!)
A great example of a Victorian crown with abundant
charm and real drama.

HHCM009
Crown Molding
1890's
Victorian

$1^1/_8$ x $4^1/_4$

Many Victorian crowns of this period were attempts
to emulate a built-up or stacked molding in a single
piece of wood—it was less expensive than multi-part
moldings. This one-piece molding looks like a bed mold
stacked on a sprung cove. An interesting example that
still works today.

HHCM010
Crown Molding
1890-1940
Victorian

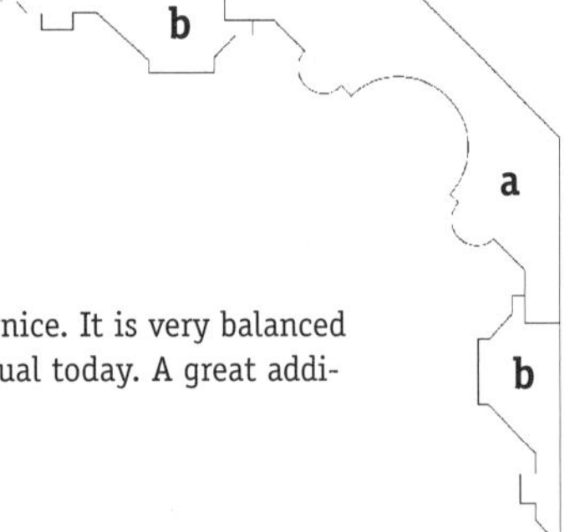

a $1^3/_8$ x $4^3/_4$
b $^7/_8$ x $2^1/_2$

Here's a Victorian built-up cornice. It is very balanced
and charming, and most unusual today. A great addition to any home.

HHCM011
Crown Molding
1890's-1940's
Classical

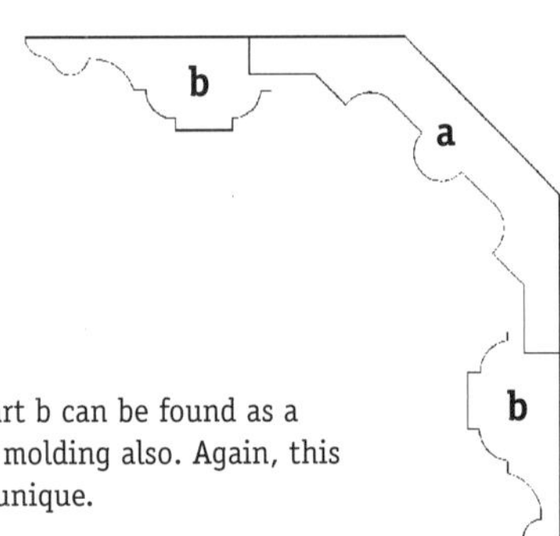

a $1^3/_8$ x $4^3/_4$
b $^7/_8$ x $2^1/_2$

Another Victorian built-up. Part b can be found as a
panel mold and a cap on base molding also. Again, this
type of mold is balanced and unique.

HHCM012
Crown Molding
1890-1920
Classical

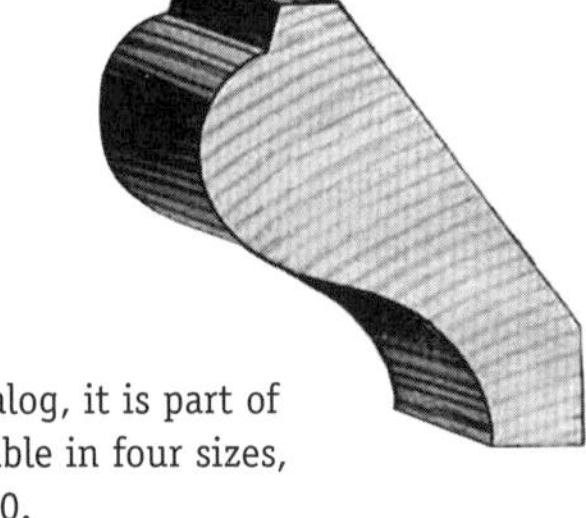

$^3/_4$ x $1^3/_4$ or $2^1/_4$ or $2^3/_4$ or $3^1/_2$

Labeled a "bed mold" in the 1890's catalog, it is part of
the Classical language. Originally available in four sizes,
it stops appearing in catalogs after 1920.

HHCM013
Crown Molding
1920-1945
Classical Revival

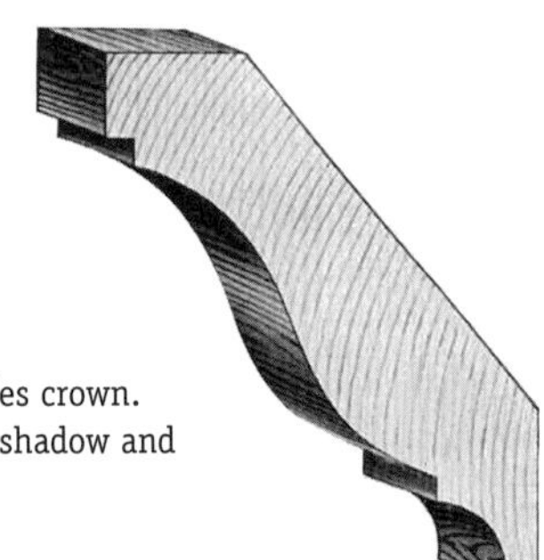

$^3/_4$ x $3^1/_2$ or $5^1/_2$

A simple twist on the classic 8000 series crown.
The wider fillet at the top adds a nice shadow and
curious effect to this mold.

HHCM014
Crown Molding
1920-1930's
Period Revival

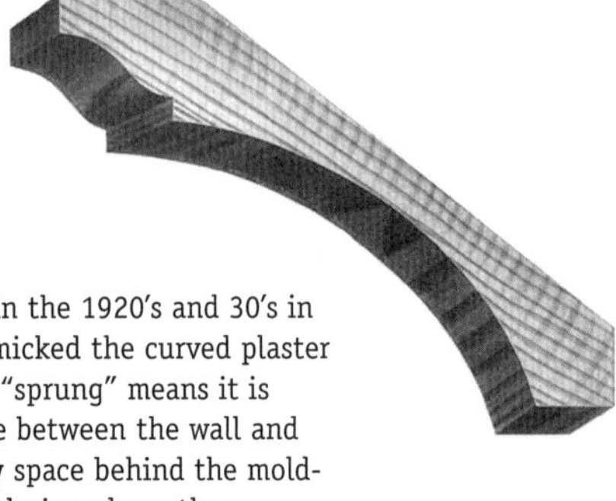

$^3/_4$ x $5^5/_8$

The sprung cove was popular in the 1920's and 30's in
Period Revival homes as it mimicked the curved plaster
used at the cornice. The term "sprung" means it is
designed to be run at an angle between the wall and
ceiling, thus leaving an empty space behind the molding. This molding adds a step design above the sprung
cove for a great look. Note also the large size.

HHCM015
Crown Molding
1925-1940
Period Revival

$^3/_4$ x 5

The odd angle used on the crown works like a sprung cove but adds extra detail. This crown is used in molding package HHMP012. Its charm was a major reason for its great popularity.

HHCM016
Crown Molding
1925-1940
Period Revival

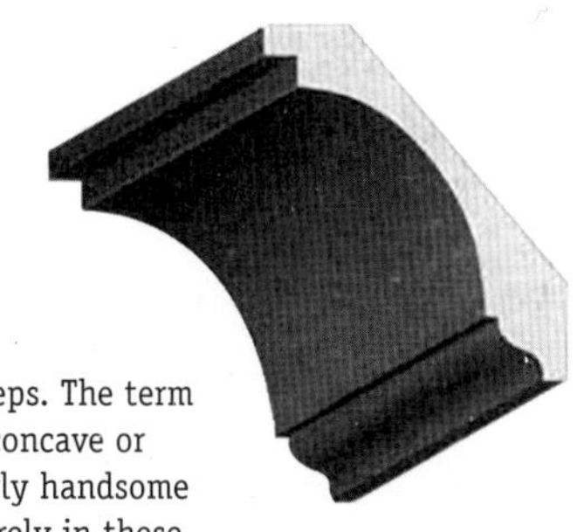

$^3/_4$ x 4$^1/_2$

Another sprung cove with added steps. The term "cove" simply means the design is concave or curved inwardly. This is a particularly handsome example still seen today, though rarely in these dimensions.

Picture Moldings

Picture molds are part of the classical architectural language, as well as serving a practical function in the home. The design of picture molds stems directly from the taenia, which was on the top of the architrave, in the entablature on a classical building (see illustration on Introduction, page 2). A picture mold functioned to protect plaster walls from damage by providing a lip for hooks to hang art. They also were frequently paired with crown moldings to add depth and richness to the room.

HHPM001
Picture Molding
1870-1890
Victorian

$^7/_8$ x 2

An early Victorian picture mold with robust details. Typical of the period.

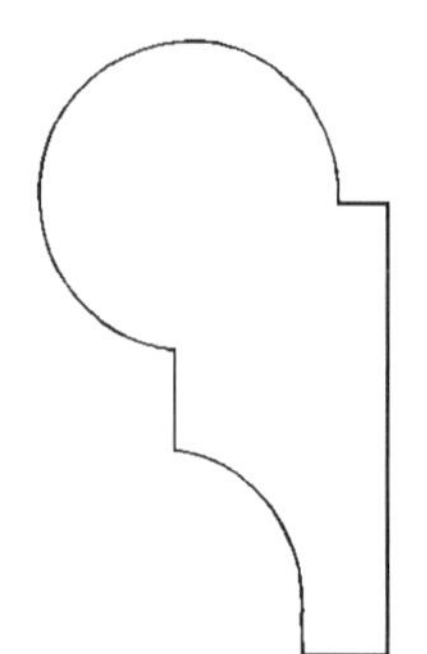

HHPM002
Picture Molding
1870-1945
Classical

$^7/_8$ x $1^3/_4$

A classic picture mold with great historical roots and lasting popularity extending from 1870 even to today.

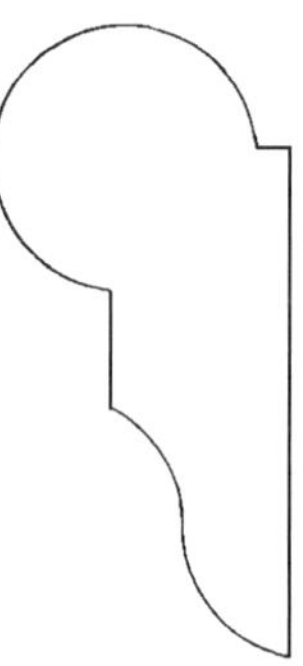

HHPM003
Picture Molding
1890's-1920's
Arts & Crafts

$^7/_8$ x 2

A simpler picture mold appropriate to Arts & Crafts style homes. The lack of adornment was a hallmark of the Arts & Crafts movement, which died out in the 1920's in the face of the Period Revival style. This also looks great stained in our custom woods.

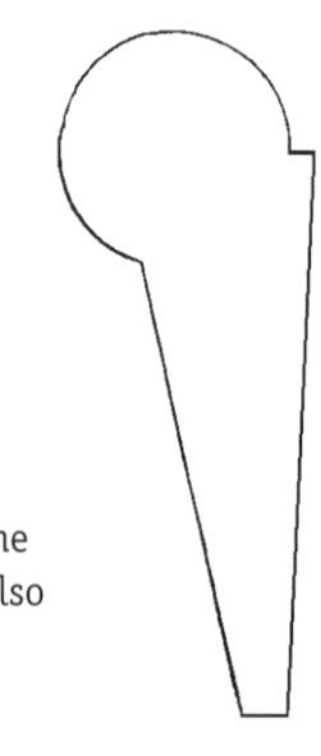

HHPM004
Picture Molding
1900-1945
Classical

$1^3/_{16}$ x $1^3/_4$

A picture mold with a long life that can still be foun with some variations today. Classical profiles makes this a popular molding.

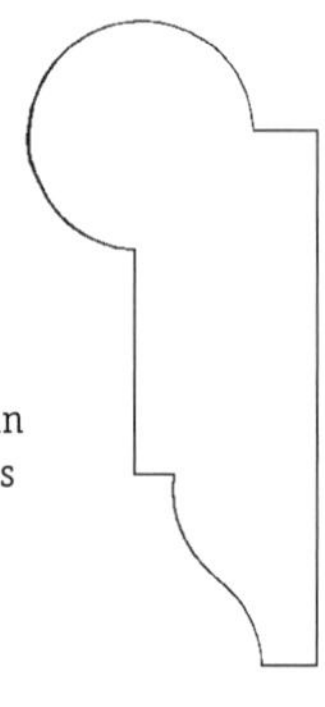

HHPM005
Picture Molding
1930's
Eclectic

$^7/_8$ x $1^5/_8$

This later-era picture mold is typical of the varied and molds that might come from a post-1930's molding catalog.

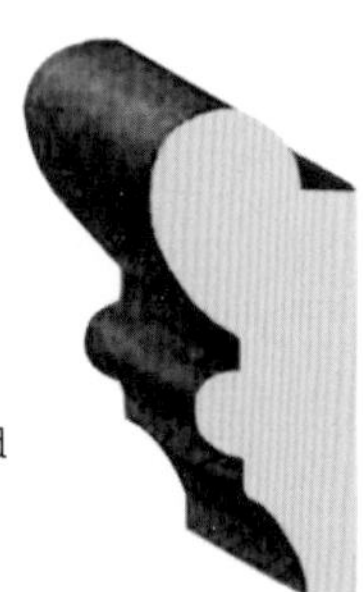

HHPM006
Picture Molding
1930's
Eclectic

$^5/_8$ x $1^3/_8$

A simple 1930's picture mold, clean and unadorned. It might be used in a modern arrangement or in historical homes of this period.

Beamed Ceilings

Beamed ceilings like this were popular in the catalogs from the turn-of-the-century through the 1930's. They were first utilized in Bungalow and Arts & Crafts style homes. This movement believed in showing the structural members of a building as they returned to a simpler look. These moldings, of course, simply emulate the look of timber framed homes, but they contributed greatly, especially when stained, to the feel and charm of these historic homes.

When ordering, remember to give your ceiling dimension and the total number of main and cross beams desired.

HHCD001

Beamed Ceiling Detail

1900-1930
Arts & Crafts/Classical

wall half beam
   3 in. projection, 6 in. drop

main beam
   5 in. wide, 5 in. drop

cross beam
   4 in. projection, 4 in. drop

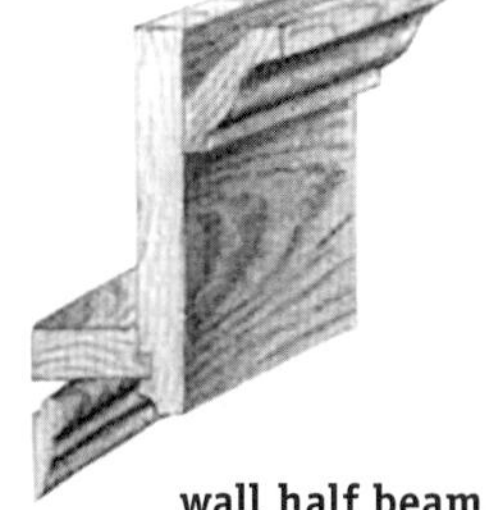

wall half beam

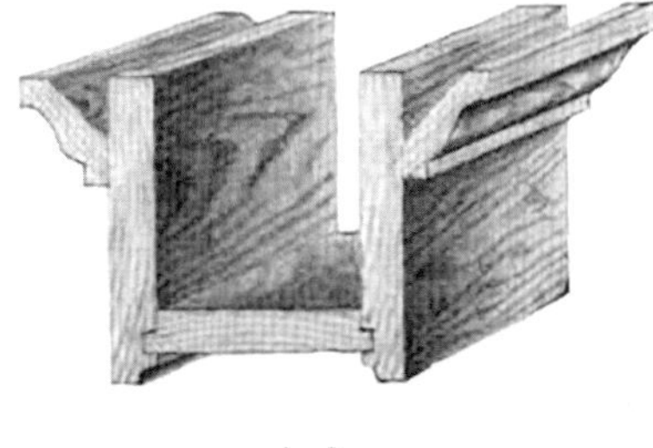

main beam

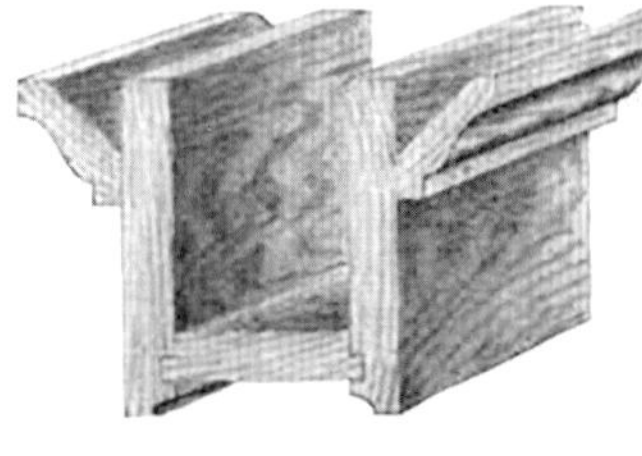

cross beam

HHCD002

Beamed Ceiling Detail

1900-1915
Arts & Crafts

wall cornice
   7 in. drop

main beam
   5 in. wide, 5 in. drop

cross beam
   4 in. projection, 4 in. drop

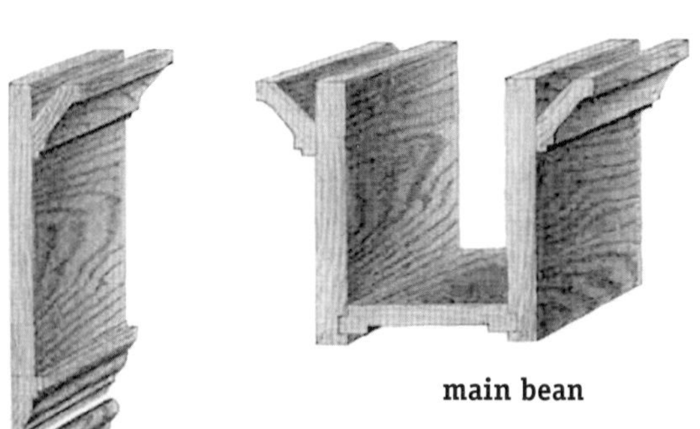

wall cornice

main bean

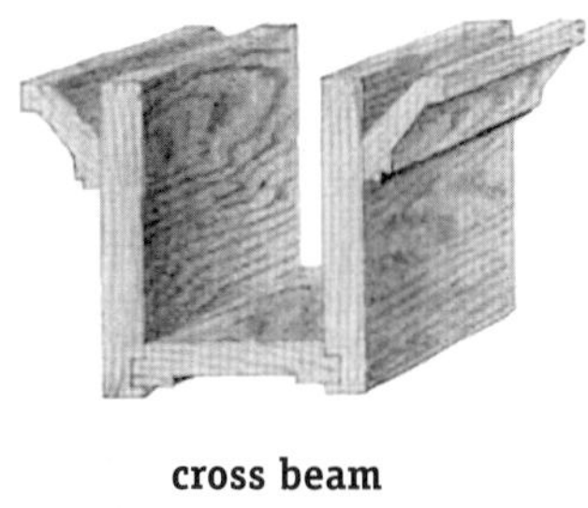

cross beam

HHCD003

Beamed Ceiling Detail

1920-1930
Classical/Period Revival

wall cornice
   3 in. projection, 3 in. drop

main beam
   6 in. wide, 3 in. drop

cross beam
   4 in. wide, 3 in. drop

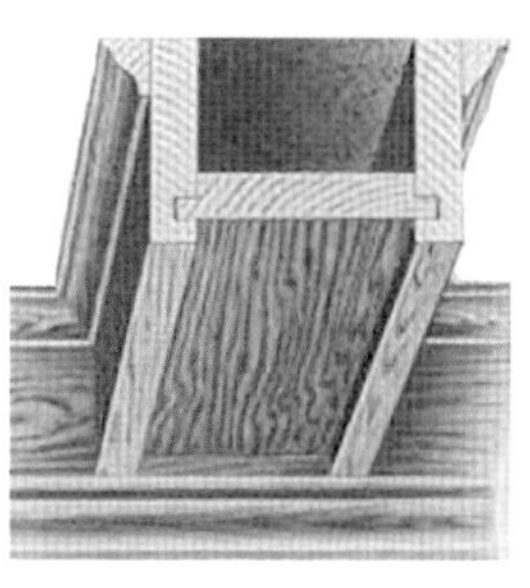

cross beam

main beam

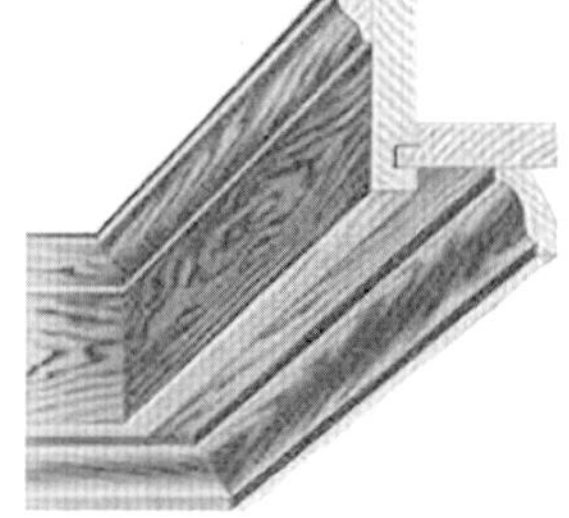

wall beam

HHCD004
Beamed Ceiling Detail

1920-1930
Arts & Crafts

wall cornice   3 in. projection, 5 in. drop
main beam   6 in. wide, 4 in. drop
cross beam   4 in. wide, 3 in. drop

Note square edge detailing, which makes this appropriate for Craftsman or Arts & Crafts style homes.

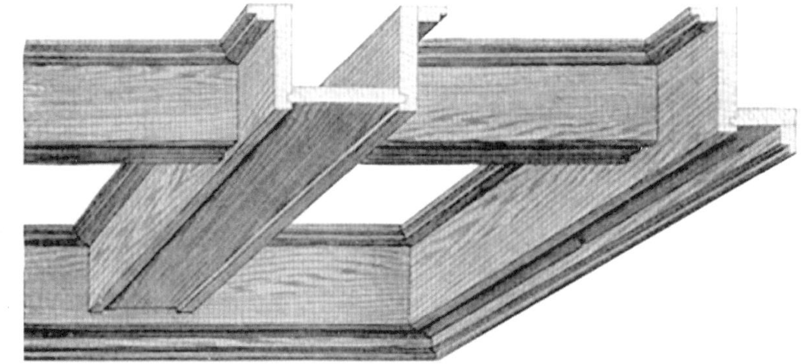

HHCD005
Beamed Ceiling Detail

1920
Period Revival

wall half beam   3 in. projection, 5 in. drop
main beam   6 in. wide, 4 in. drop
cross beam   4 in. wide, 3 in. drop

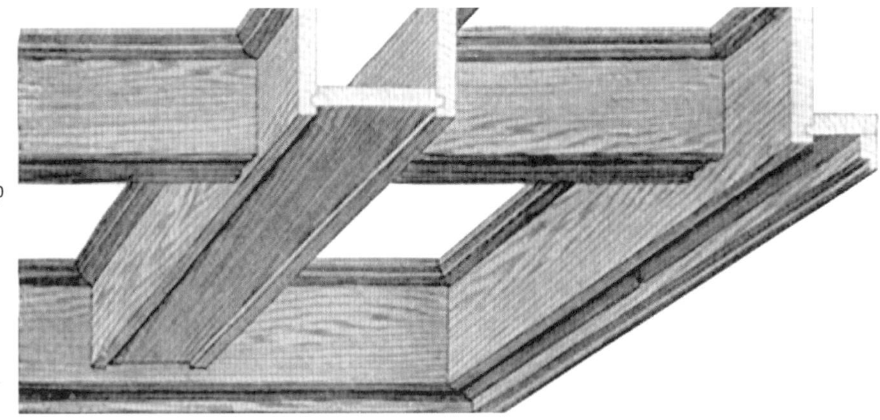

HHCD006
Beamed Ceiling Detail

1920
Period Revival

wall half beam
   1 in. projection, 5 in. drop

cross beam
   4 in. wide, 3 in. drop

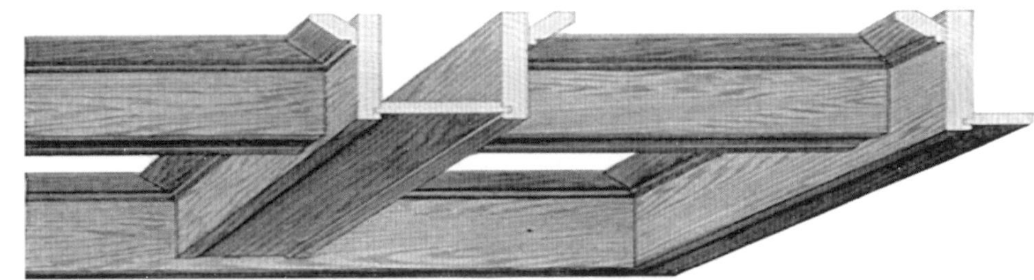

HHCD007
Beamed Ceiling Detail

1920
Arts & Crafts

wall half beam
   3 in. projection, 3 in. drop

main beam
   8 in. wide, 3 in. drop

cross beam
   4 in. wide, 3 in. drop

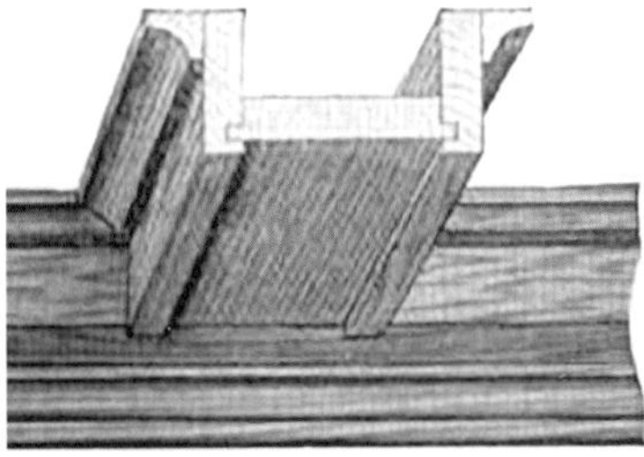

cross beam

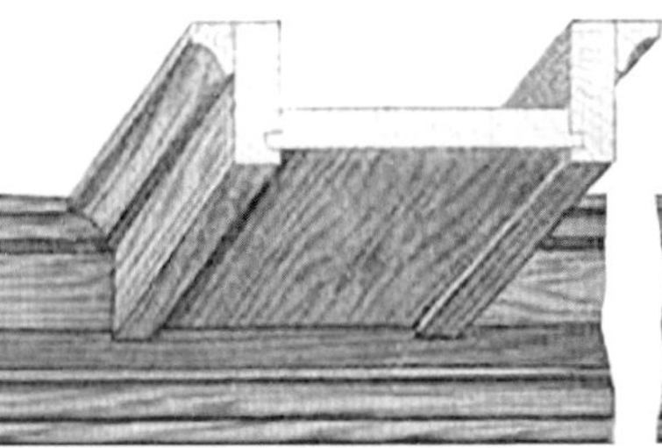

main beam

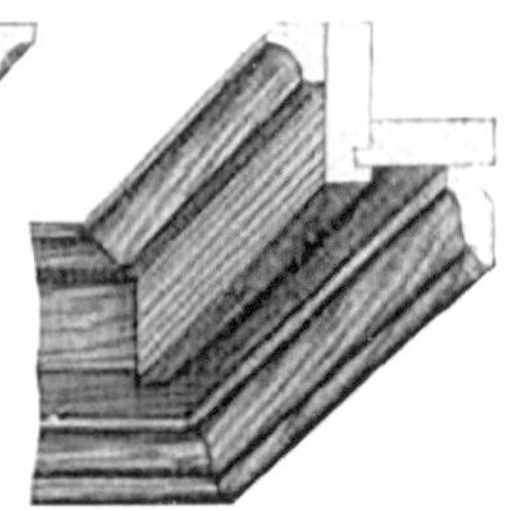

wall beam

HHCD008
Period Cornice

1925-1935
Period Revival

a   HHCM015
b   HHPM002

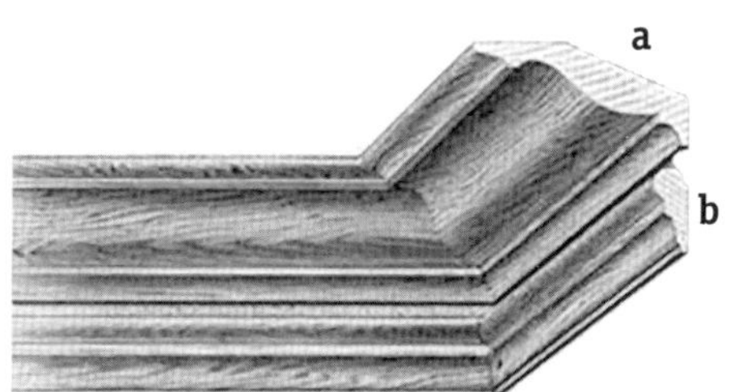

HHCD009
Period Cornice

1925-1935
Period Revival

a   HHCS048
b   HHCM014
c   HHPM002

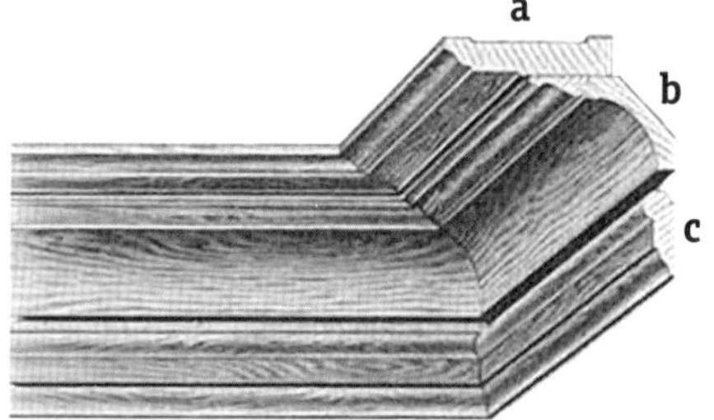

HHCD010
Period Cornice

1925-1935
Period Revival

a   HHPA004
b   HHCM004
c   HHPM002

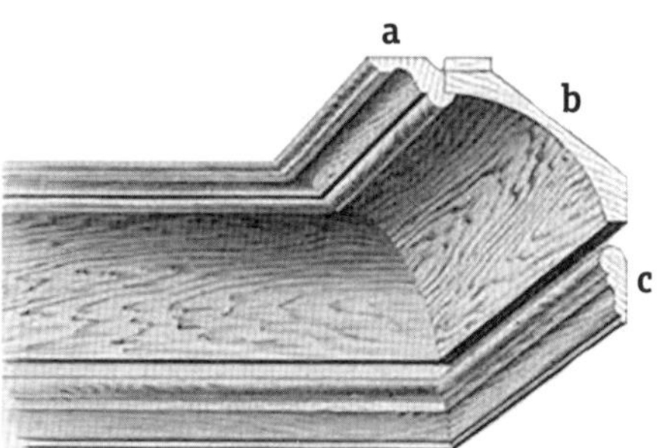

HHCD011
Period Cornice

1920-1935
Period Revival

call for details

This is a great detail that gives the ceiling a paneled look.

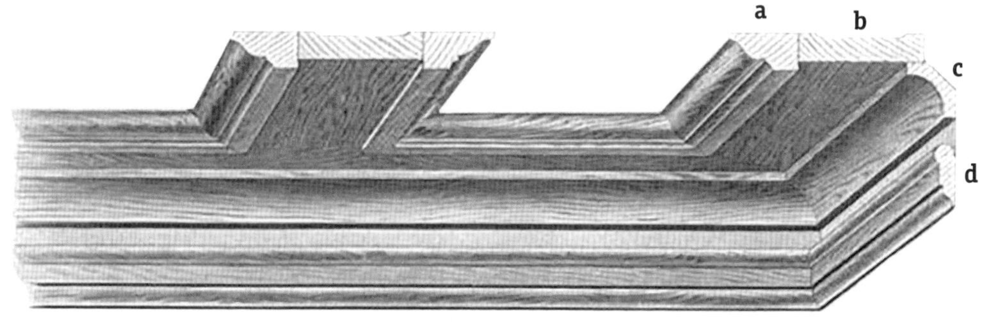

BASE, BASE CAP, SHOE MOLDS AND PLINTH BLOCKS

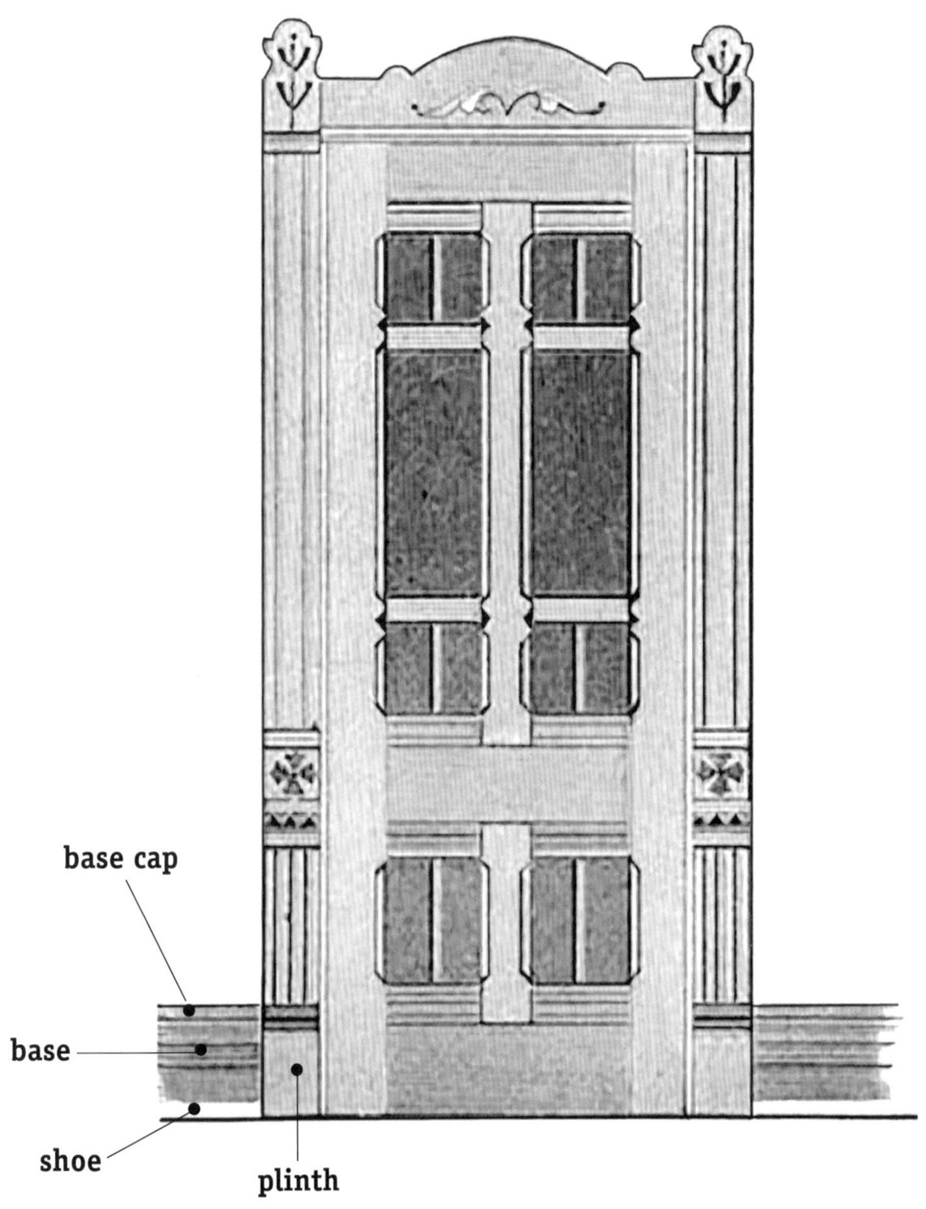

Base Moldings

Surprisingly, there are only 15 typical base moldings found in catalogs between 1870 and 1940. Designers, however, utilized base caps (which sit on the top of the base molding) and shoe molds (which sit at that bottom) to create a great variety of designs. We show moldings with base caps as they appeared in the original catalogs. You can achieve a greater variety of looks by selecting from the different base cap and shoe molds offered here in this catalog.

A major change we see in the standard catalogs is the height of the moldings. Victorian rooms were typically larger and taller than rooms in later styles. For this reason, Victorian moldings are taller than what we see later. For example, Victorian base moldings are normally 7-10 inches high, with a base between 7-8 inches high and a base cap of 2-3 inches. (You will notice that some Victorian bases are lipped; this design was intended to accept a base cap.) The smaller Arts & Crafts rooms utilized a 5-6 inch high base during the 1920's. By the 1930's, the standard height of a base molding was 3-4 inches.

For this reason, we offer the moldings that were available in every period in 3-inch, 5-inch, and 7-inch heights. Others are listed in one size, indicating they only appeared in that applicable time period.

HHBM001
Base

1890-1920
Classical

$^3/_4$ x 5 $^1/_2$ or 7 $^1/_2$

A classic molding that is ready to accept a base cap. Available in the Victorian era and up to the 20's, it is versatile in style and can work without a base cap in an Arts & Crafts style home.

HHBM002
Base

1890-1915
Victorian

$^3/_4$ x 7 $^1/_2$

A Victorian base mold that was a popular treatment. It is a simpler base mold because it would not accept a base cap. The top detail is meant to mimic the cap that might appear on a more ornate base.

HHBM003
Base

1890-1915
Victorian

$^3/_4$ x 7 $^1/_2$

Another Victorian base similar to HHBM002 but with a beaded detail two-thirds of the way up. This detail looks great painted or stained and is typical of the ornate Victorian moldings at the time.

HHBM004
Base

1890's
Victorian

$^3/_4$ x 7 $^1/_2$

A short-lived Victorian base ready to accept a base cap. The heavy beaded detail showed up in the first Universal Molding Catalog but was cut after the first addition.

HHBM005

Base

1890's
Victorian

$^3/_4$ x 7 $^1/_2$

A handsome Victorian base ready to accept a base cap. A little subtler than other Victorian bases, it still carries the height to work well in any Victorian room.

HHBM006

Base

1890-1940

$^7/_8$ x 4 $^1/_2$ or 5 $^1/_2$ or 7 $^1/_2$

The flat casing like this one was available in all historic catalogs, but it was used more in the Arts & Crafts style home. Because of its flat, plain simplicity it worked especially well when stained. This molding's diversity allows it to work well in new homes today, especially with the right base cap.

HHBM007

Base

1890's-1940's
Classical

$^3/_4$ x 3 $^1/_2$ or 5 $^1/_2$ or 7 $^1/_2$

This classic molding had an extended life, smaller-scale versions are still seen in modern homes. The top is meant to mimic a base cap, but this simpler one-piece molding works in many homes of various styles.

HHBM008

Base

1890's
Victorian

$^3/_4$ x 7 $^1/_2$

The last of the Victorians. it is also meant to mimic a base with base cap. It is a mold that would work well in simpler Victorian buildings.

HHBM009

Base

1910-1945
Classical

$^3/_4$ x 3 $^1/_2$ or 5 $^1/_2$ or 7 $^1/_2$

Similar to HHBM006 but with an eased or curved edge. This eased edge detail was used commonly in the Arts & Crafts period, though it found use beyond this era. A simple and subtle statement that would work well stained or painted.

HHBM010

Base

1890-1915
Classical

$^3/_4$ x 3 $^1/_2$ or 5 $^1/_2$ or 7 $^1/_2$

Another classical base molding. In Victorian times we see the taller version. The smaller sizes come into vogue later, but the bead and cove detail at the top is timeless.

HHBM011

Base

1890-1945
Classical

$^3/_4$ x 3 $^1/_2$ or 5 $^1/_2$ or 7 $^1/_2$

Similar to HHBM010 but with an OG detail instead of the bead and cove. This molding would be common in simpler homes where they couldn't afford a second base cap.

HHBM012

Base

1920
Arts & Crafts

$^3/_4$ x 5 $^1/_2$

This square-edged base was an Arts & Crafts design detail that works well in homes of simple styling. It looks great stained, but we think the shadow contrast is best achieved when painted.

HHBM013
Base
1920-1930
Period Revival

$^{3}/_{4}$ x 3$^{1}/_{2}$ or 5$^{1}/_{2}$

A great Period Revival molding. It is similar to HHBM010,
and yet it doesn't show up in the catalogs until 1920.
A great all-around molding that can work well in old and
new homes alike.

HHBM014
Base
1920-1945
Arts & Crafts

$^{3}/_{4}$ x 3$^{1}/_{2}$ or 5$^{1}/_{2}$

An Arts & Crafts molding with Victorian roots that lasts
through to the 40's. Similar to HHBM012 but with a simpler
top design, it works well in many situations. It can also accept
a base cap if desired.

HHBM015
Base
1925-1945
Eclectic

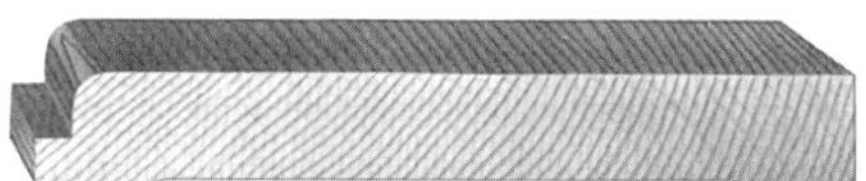

$^{3}/_{4}$ x 3$^{1}/_{2}$ or 5$^{1}/_{2}$

An Eclectic base mold that surfaces in the Period Revival
era. It doesn't fit into any clear style, but this adds to its
diversity. It could even work with base cap if desired.

Base Caps

Base caps and panel molds were not separated in the historical catalogs, so if you don't find a base cap you like here, try the panel molds section. If you are familiar with new construction you may have seen plans calling for a panel mold called a PM5 to be placed on top of a base to create an historic look.

You won't find the PM5 in our book because it is not historic. (It is "historic-esque," but don't get me started on all the modern products masquerading as historic!) We sell only authentic historical moldings in exactly the same shape as they were originally offered. We're confident that once you experience the real deal, you will accept no substitutes.

The base caps here are all rabbeted, which makes for an easier installation in most cases. If you want a simpler mold to sit on top of the base, try HHPA001 or HHPA002.

HHBC001
Base Cap Molding
1880-1900

$1^1/_8$ x 3

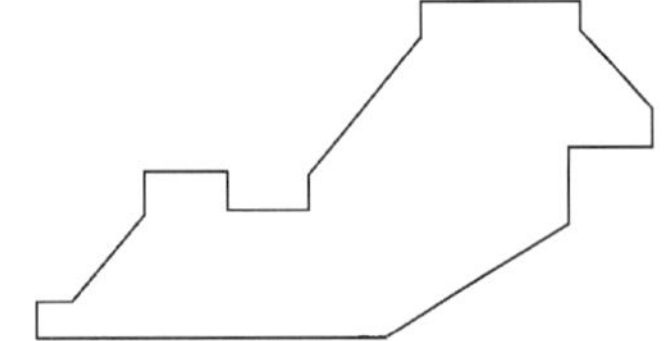

This base cap is nice and angular and would really dress up a base mold in an historic house. This is mean to fit on a $^3/_4$-inch base.

HHBC002
Base Cap Molding
1880-1915

$1^1/_8$ x $2^3/_4$

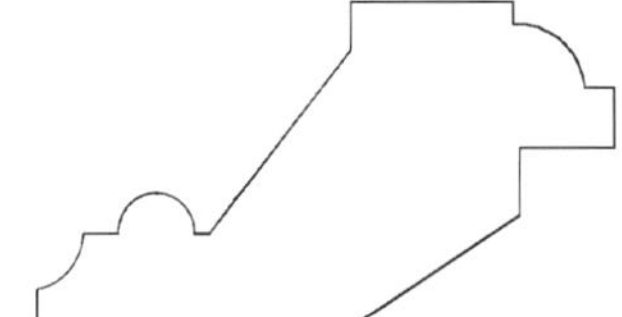

Another base cap, similar to HHBC003, but with a cove on the end in a classic fashion.

HHBC003
Base Cap Molding
1880-1900

$1^1/_8$ x $2^1/_2$

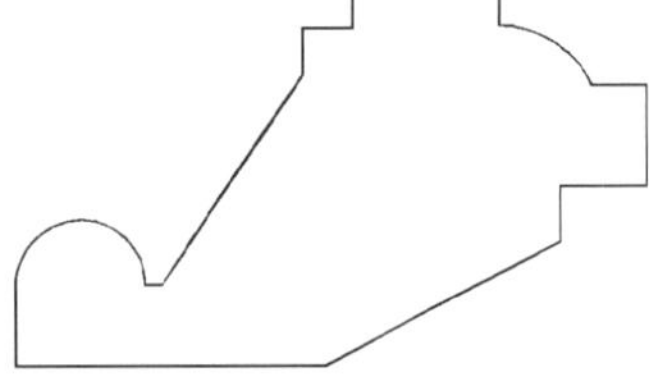

Again similar to HHBC002, but this one is more compact, perfect for smaller rooms that can't afford the height of a typical Victorian cap. If your rooms have 8-foot ceilings, this shorter version could be just the cap you need.

HHBC004
Base Cap Molding
1870-1925

$^7/_8$ x $2^1/_8$

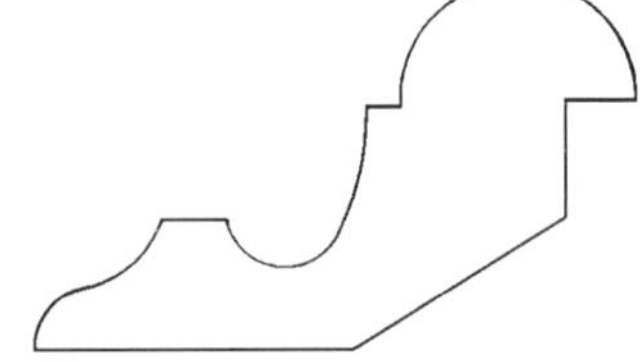

The lines of this base cap are classic and outlive the Victorian era, finding use for many years afterwards. This is base cap is one of my personal favorites.

Shoe Moldings

Modern shoe molds are boring! Historically, however, there was greater diversity. Especially in Victorian times, shoe moldings were quite ornate and attractive. We offer six types of historical shoe moldings.

HHSM001
Shoe Molding

1900-1945
Classical

$^3/_4$ or $^7/_8$ or 1 inch

A quarter round shoe mold was a common historical treatment. Its use declines after 1910. The taller Victorian bases used the $^7/_8$" or 1" quarter round for better scale.

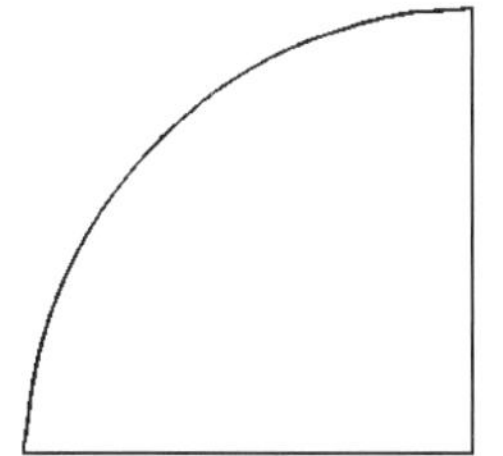

HHSM002
Shoe Molding

1890-1915
Victorian

$^3/_4$ x 1

A Victorian shoe mold that is seen in catalogs up through 1915. It is taller (at 1 inch) than normal shoes, so it needs to be used on a larger base molding. Attention to details like this can add real value to any project.

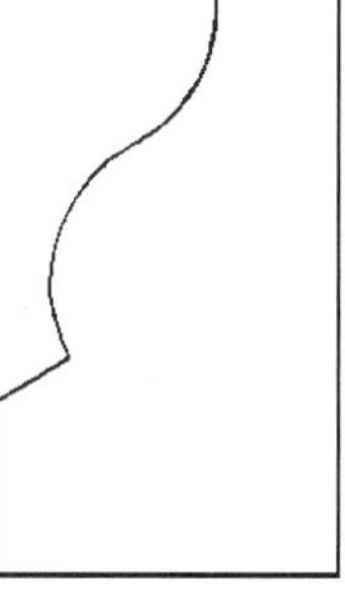

HHSM003
Shoe Molding

1890-1900
Victorian

$^3/_4$ x 1

A popular Victorian shoe with an OG profile, it would go well on any pre-1900 home.

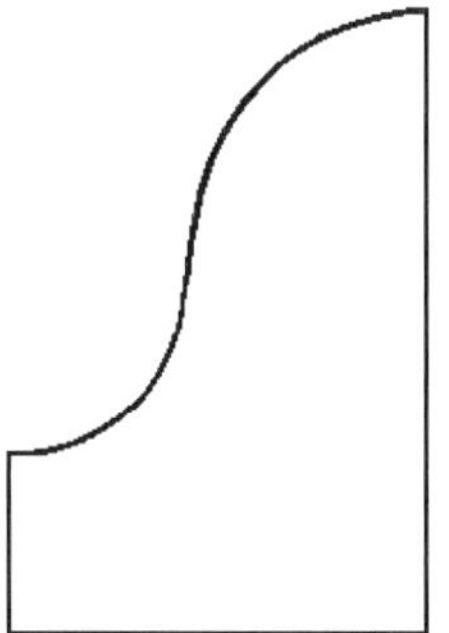

HHSM004
Shoe Molding

1890-1900
Victorian

$^7/_8$ x $^7/_8$

Another distinctive Victorian shoe mold, it is quite wide at the base and was popular in the 1880's and 1890's. Because of the shape of the profile it can even work on Gothic or Greek Revival styles as it uses an elliptical radius typical of these styles.

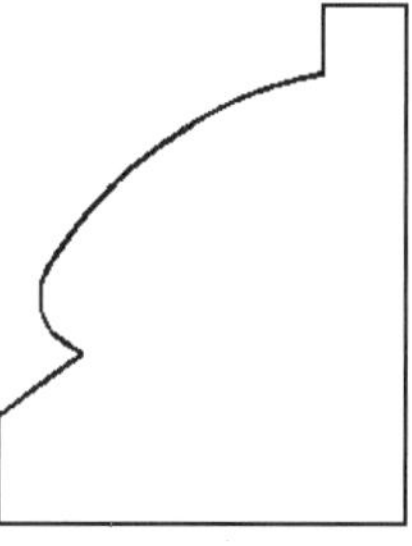

HHSM005
Shoe Molding

1900-1940
Classical

$^7/_8$ x $^7/_8$

A later shoe mold that lasts up into the 40's. We used it on a 1920's historical building we restored in Houston. It was the first time I had seen it outside of the catalog it and provided a nice detail at the floor.

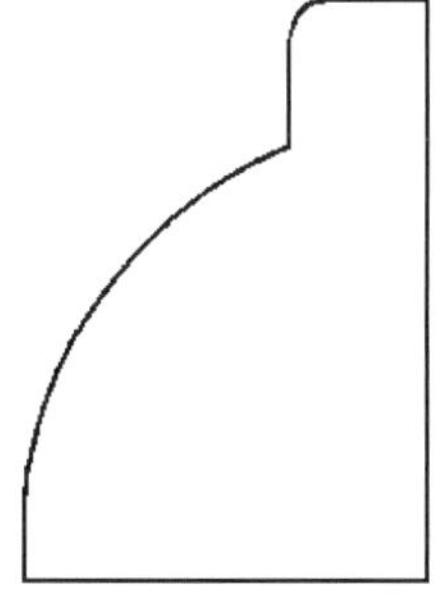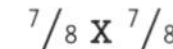

HHSM006
Shoe Molding

1920
Arts & Crafts

$^7/_8$ x $^7/_8$

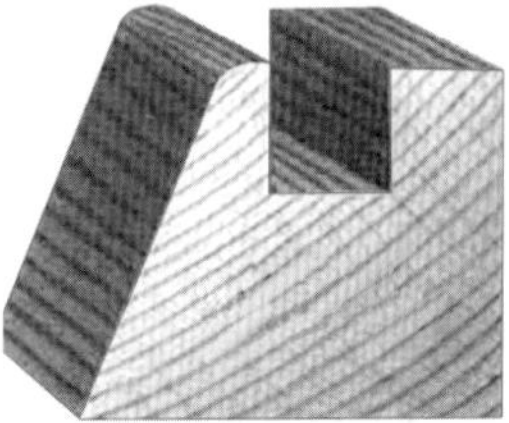

A wonderful shoe that was not seen in catalogs until the 1920's. It works well and was paired with HHBM014 in the original catalogs. Let us know if you are going to using these two together so we can cut a slot in the base. This slot adds functionality, allowing you to hide large gaps between floor and wall.

Plinth Blocks

The plinth block is a decorative trim piece at the bottom corners of a doorway. There are 19 plinth blocks listed here, although the variety of plinths blocks you will see in the field is much greater than this. Many mills offered their own plinth block designs, which differed from region to region and did not appear in the standard catalogs.

Plinths are part of the language of classical architecture. They represent the lowest portion of the column at the base and usually signify the flat portion of the base. This again demonstrates classical architecture's strong influence in historical molding design.

The decorative Victorian plinth (the first 18 models shown) lasts in catalogs up to 1915. It is built up and bigger in size to complement the built up Victorian details in base, casing or door. The square block version (HHPB0019) endured in molding catalogs into the 1930's.

HHPB001
Plinth Block
1885-1915
Victorian

$1^3/_8$ thick x $5^1/_2$ w x 12h

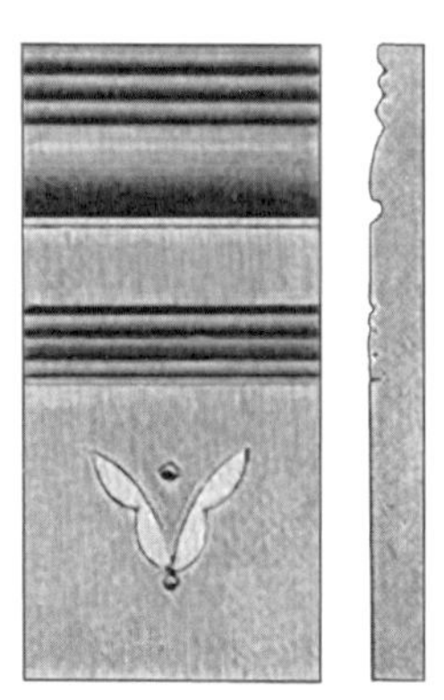

HHPB002
Plinth Block
1885-1915
Victorian

$1^3/_8$ thick x $5^1/_2$ w x 12h

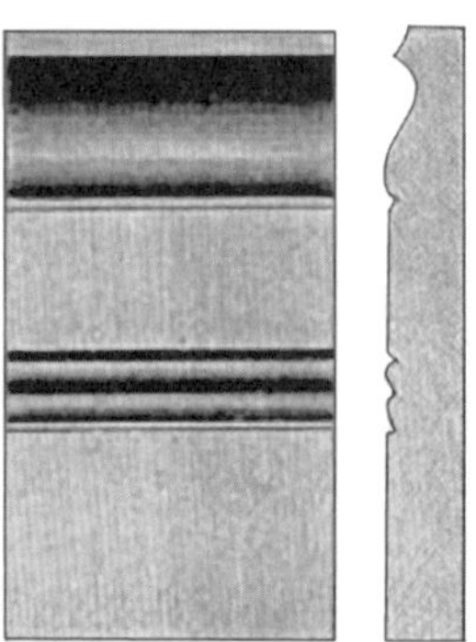

HHPB003
Plinth Block
1885-1915
Victorian

$1^3/_8$ thick x $5^1/_2$ w x 12h

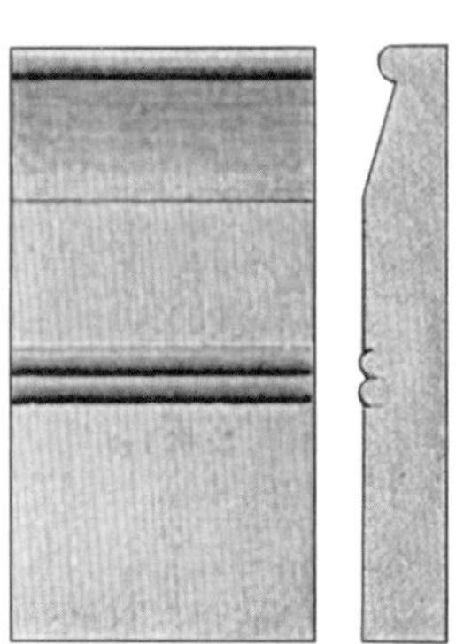

HHPB004
Plinth Block
1885-1915
Victorian

$1^3/_8$ thick x $5^1/_2$ w x 12h

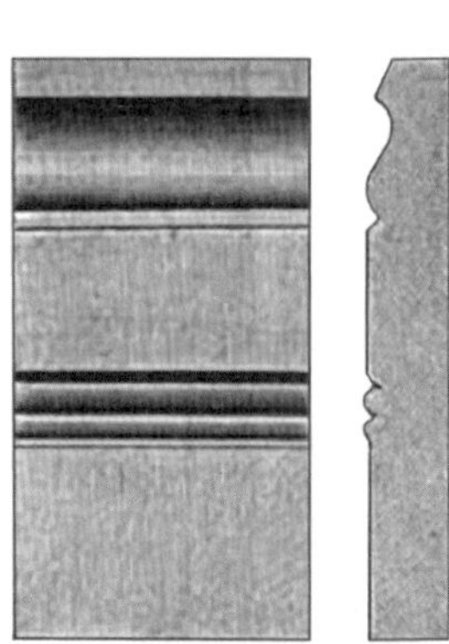

HHPB005
Plinth Block
1885-1915
Victorian

$1^3/_8$ thick x $5^1/_2$ w x 12h

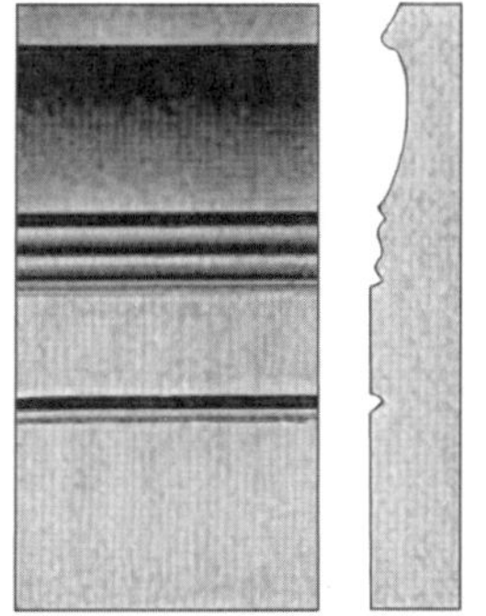

HHPB006
Plinth Block
1885-1915
Victorian

$1^3/_8$ thick x $5^1/_2$ w x 12h

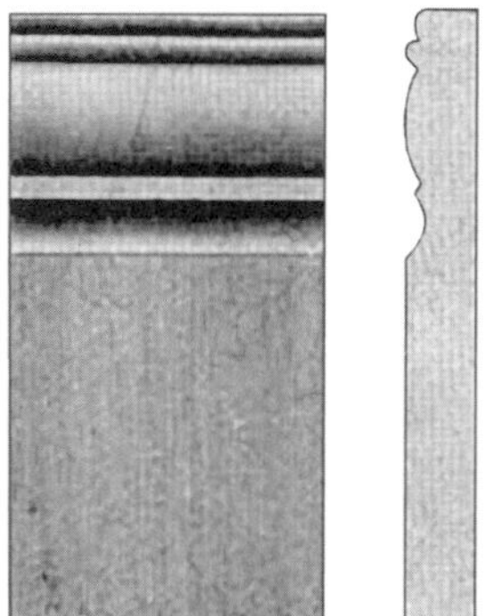

HHPB007
Plinth Block
1885-1915
Victorian

$1^3/_8$ thick x $5^1/_2$ w x 12h

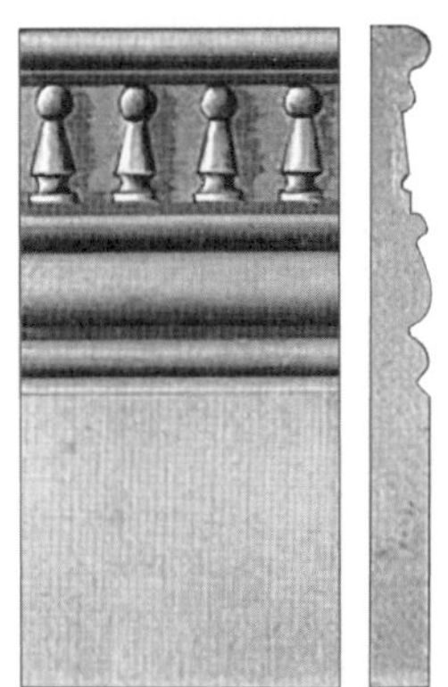

HHPB008
Plinth Block
1885-1915
Victorian

$1^3/_8$ thick x $5^1/_2$ w x 12h

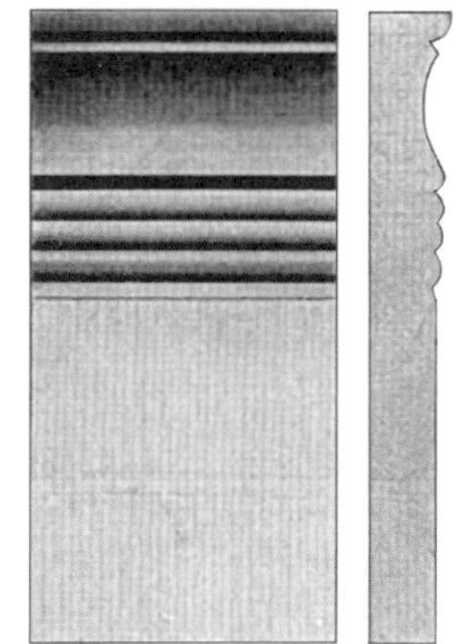

HHPB009
Plinth Block
1885-1915
Victorian

$1^3/_8$ thick x $5^1/_2$ w x 12h

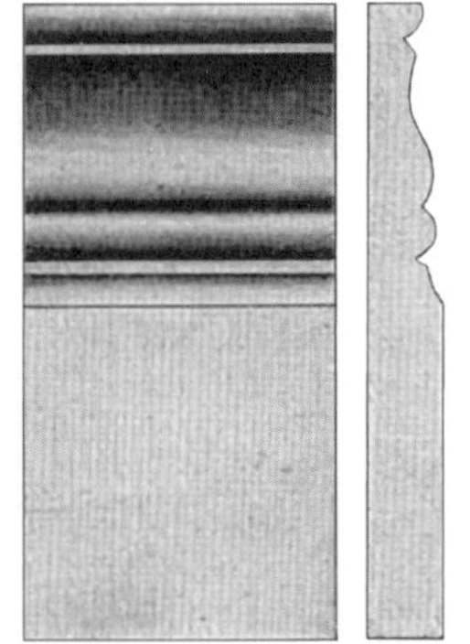

HHPB010
Plinth Block
1885-1915
Victorian

$1^3/_8$ thick x $5^1/_2$ w x 12h

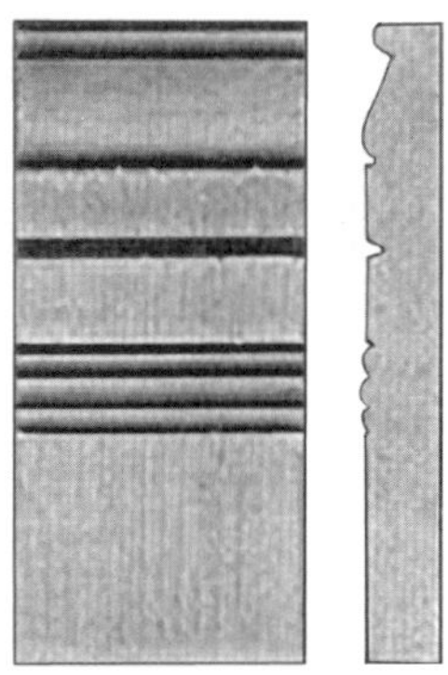

HHPB011
Plinth Block
1885-1915
Victorian

$1^3/_8$ thick x $5^1/_2$ w x 12h

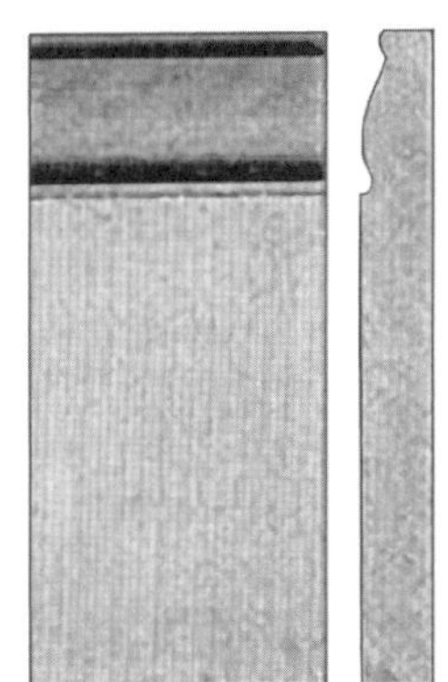

HHPB012
Plinth Block
1885-1915
Victorian

$1^3/_8$ thick x $5^1/_2$ w x 12h

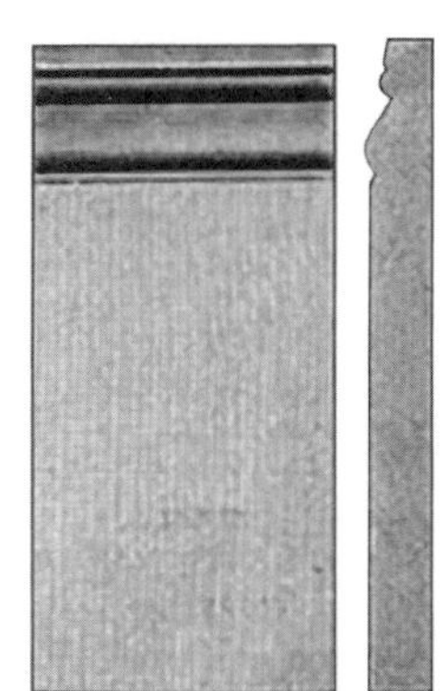

HHPB013
Plinth Block
1885-1915
Victorian

1³/₈ thick x 5¹/₂ w x 12h

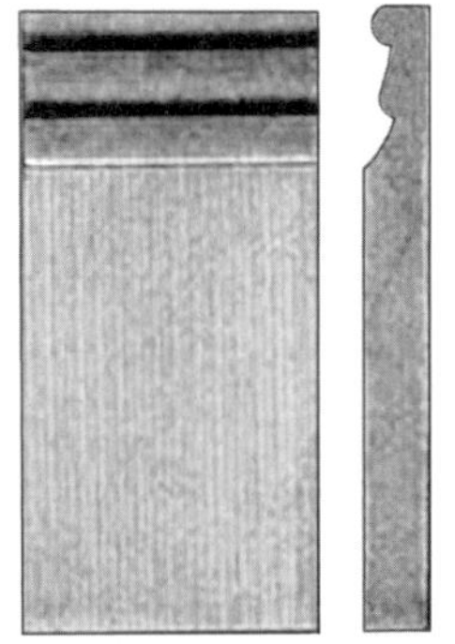

HHPB014
Plinth Block
1885-1915
Victorian

1³/₈ thick x 5¹/₂ w x 12h

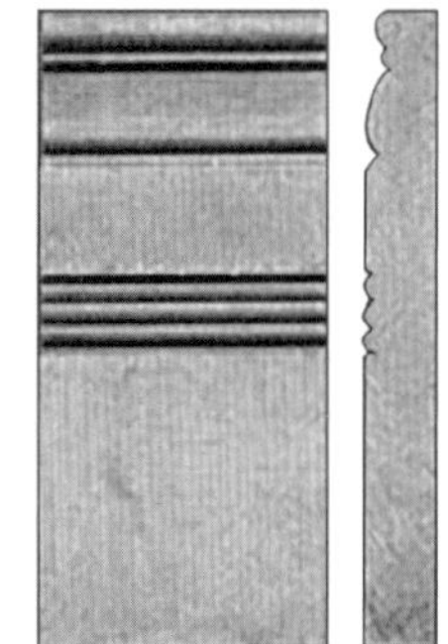

HHPB015
Plinth Block
1885-1915
Victorian

1³/₈ thick x 5¹/₂ w x 12h

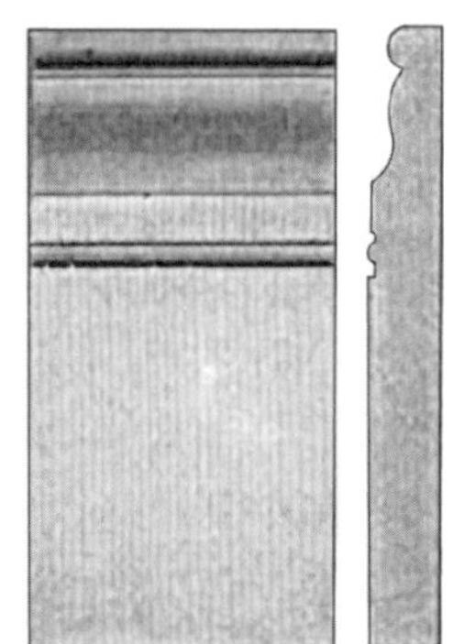

HHPB016
Plinth Block
1885-1915
Victorian

1³/₈ thick x 5¹/₂ w x 12h

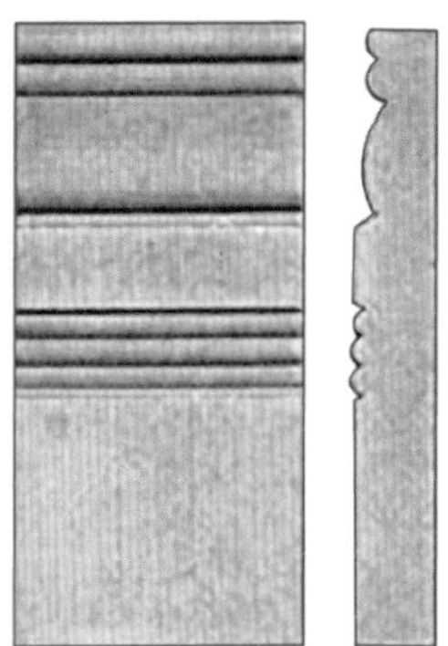

HHPB017
Plinth Block
1885-1915
Victorian

1³/₈ thick x 5¹/₂ w x 12h

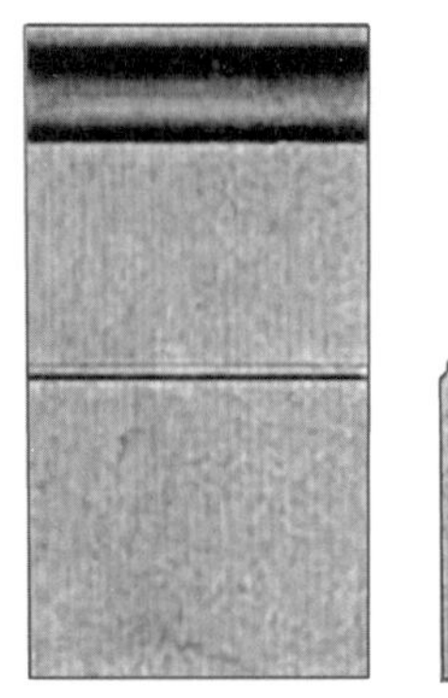

HHPB018
Plinth Block
1885-1915
Victorian

1³/₈ thick x 5¹/₂ w x 12h

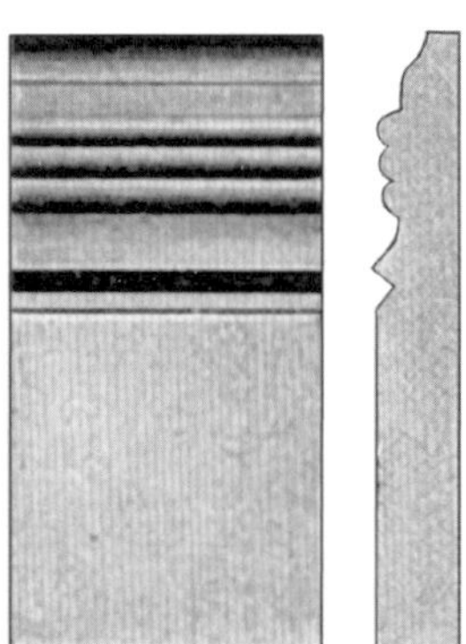

HHPB019
Plinth Block
1900-1930
Arts & Crafts

1³/₈ thick x 5¹/₂ w x 12h

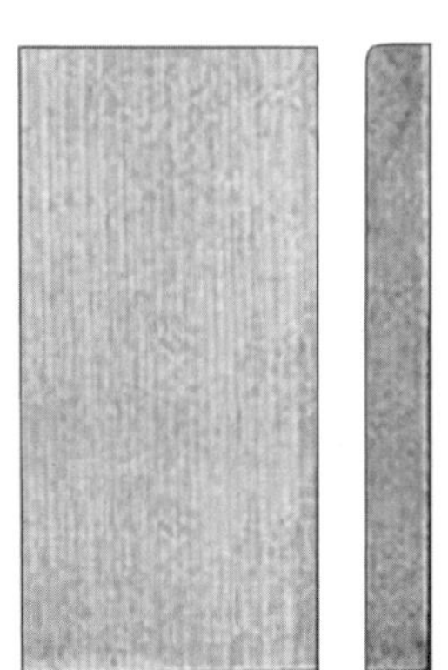

Casings

CASINGS, BAND MOLDINGS, BACK BANDS, STOPS, WINDOW STOOLS, ROSETTES, CORNER BLOCKS AND HEADER BLOCKS

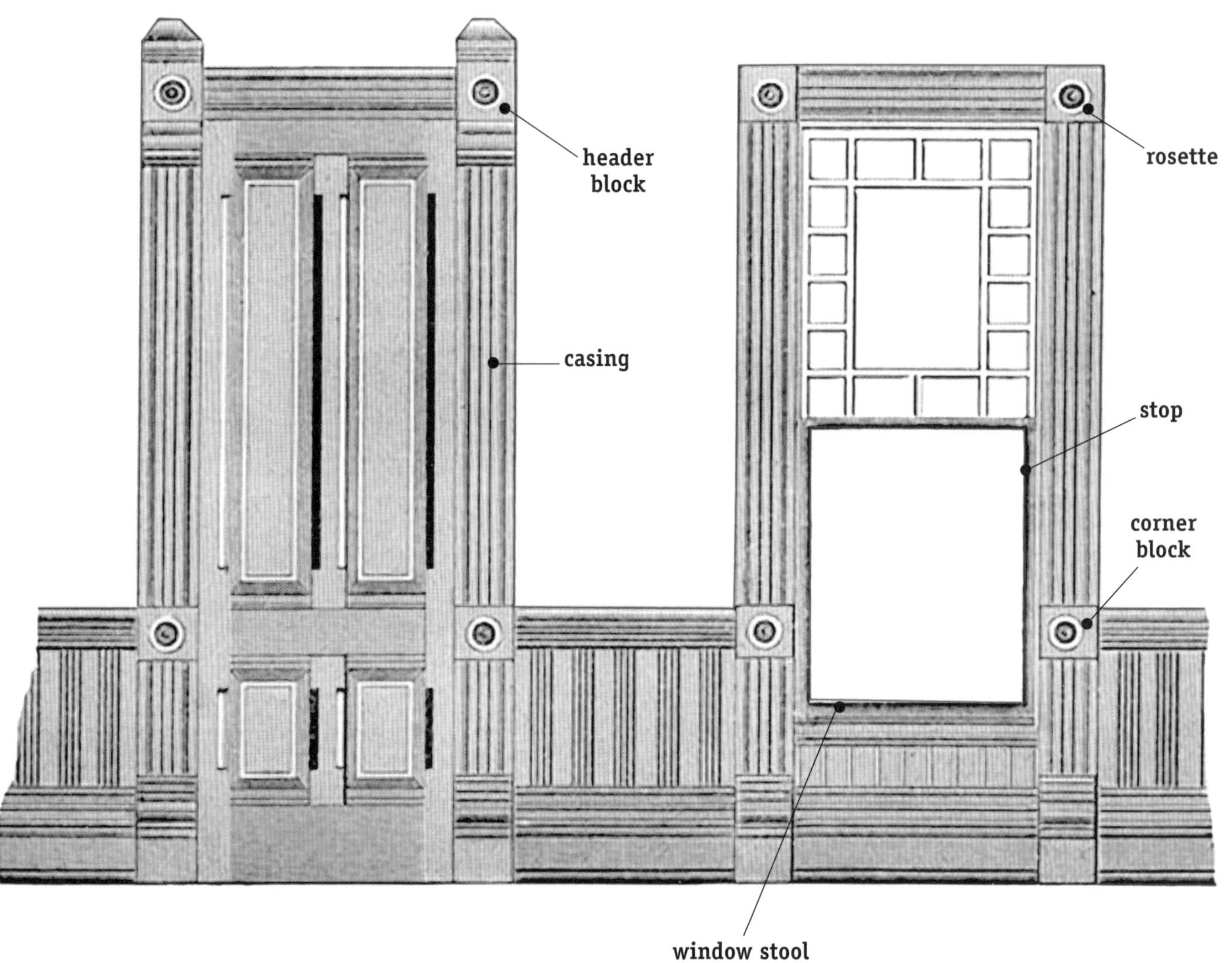

HULL
HISTORICAL
MOLDING

Casings

Try not to get overwhelmed with the number of casings listed in this section. There are, in fact, more casings than any other molding type found in this catalog. The main culprit in this explosion of choice is the Victorian period, as the standard catalogs of that era were filled with casings.

Victorian casings are often symmetrical in design. In symmetrical casings, the design on the left side matches the design on the right. Casing HHCS003, for example, is symmetrical, while casing HHCS002 is not. Symmetrical casings are usually paired with rosettes and header blocks. Other casings have a profile on the front end only. The square back edge permits the molding to be paired with a back band, and also allowed the molding to be mitred at the door or window corners rather than requiring a rosette or corner block.

You will notice that the moldings generally grow smaller from 1890 coming forward. Rooms are smaller and the scale of moldings shrinks as well. If you find a casing design you like but need it in a different size, let us know and we can custom run it in the size you need.

The back bands and band molds shown on some of these moldings are as they appeared in the original catalogs. The pairings they show usually reflect the most popular examples from that period. In many cases they are the best match, but feel free to also pair other back bands or band molds with your casings to achieve the look you want.

Casings are a great part of any molding package. Enjoy ours and let us know if you need help putting packages together. The brief descriptions may help and you can also look at the Molding Packages section if you want a turnkey design look. We are also available to consult on your project and provide the right moldings for new or historic homes.

HHCS001
Casing
1870-1915
Victorian

$3/4$ x $5^1/2$

This early Victorian casing was very simple and often used with a back band or band molding. Because of its simplicity it can work on many styles of homes today.

HHCS002
Casing
1890's
Victorian

$3/4$ x $4^1/2$

Another of the simple Victorian casings that was used with a back band or band molding. The square edged detailing would work well with many contemporary homes.

HHCS003
Casing
1890's
Victorian

$3/4$ x 5

A classical symmetrical casing, very common in the Victorian era. These were typically paired with rosettes and header blocks. Rarely do we see them simply mitred together at the corner.

HHCS004
Casing
1890's
Victorian

$3/4$ x 5

A simpler Victorian cove casing, it is still symmetrical and there are enough angles and sweeps to provide a nice shadow when painted.

HHCS005
Casing

1890's
Victorian

$^3/_4$ x 5

A very interesting Victorian casing that is only seen in the 1890's catalog. The square edged fillets flanking the oversized bead are unique and not a common detail. There are many sharp edges on this molding, which would look great painted or stained.

HHCS006
Casing

1890's
Victorian

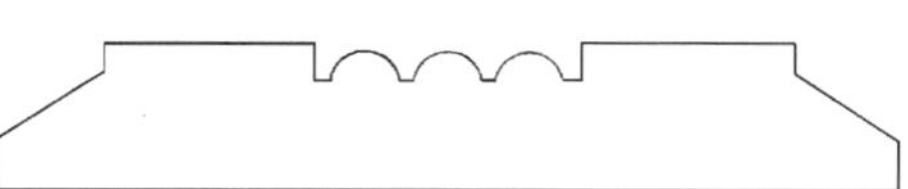

$^3/_4$ x 5$^1/_2$

The reed or beaded detail on this molding is a Victorian pattern that is seen over and over again. This symmetrical molding was most frequently used with rosettes and plinths.

HHCS007
Casing

1890-1914
Victorian

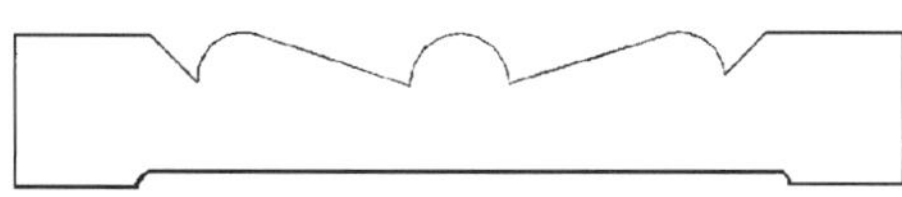

$^3/_4$ x 5$^1/_4$

A Victorian casing that we have used on a number of jobs. It was somewhat popular and lasts in the catalogs until the 1914 edition. The center bead is quite handsome and symmetrical and really sets off a door or window with subtle grace.

HHCS008
Casing

1890's
Victorian

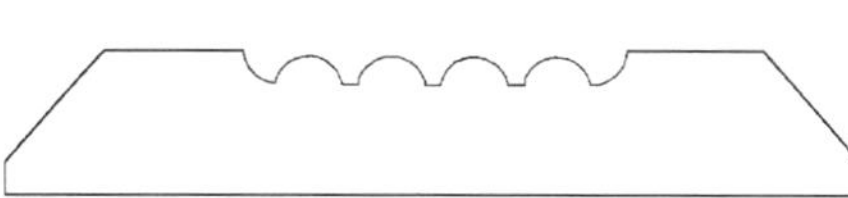

$^3/_4$ x 5

Another reed or bead pattern that is ornate and unique. It makes a very historic and interesting statement in any home.

HHCS009
Eastlake Casing

1890's
Victorian

$^3/_4$ x 4$^1/_2$

Charles Eastlake was an English designer who gained prominence in England and America in the 1880s and 1890s. His book Hints on Household Taste was widely read. His name was attributed and attached to many home styles and period details that he never actually designed. This "Eastlake" casing is an example of that product.

HHCS010
Eastlake Casing

1890's
Victorian

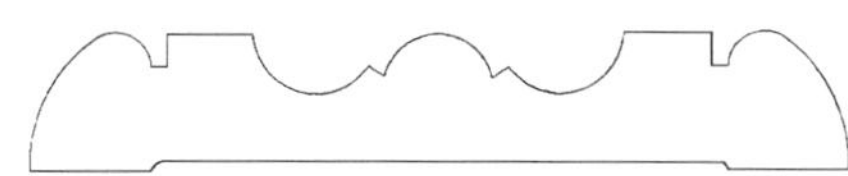

$^3/_4$ x 5$^1/_2$

Another Eastlake design (see HHCS009) that is a real eye-catcher. This does not have a long life in the catalogs but it was a typical Victorian mold. It is very bold and symmetrical and would work with any ornate rosette or plinth.

HHCS011
Eastlake Casing

1890's
Victorian

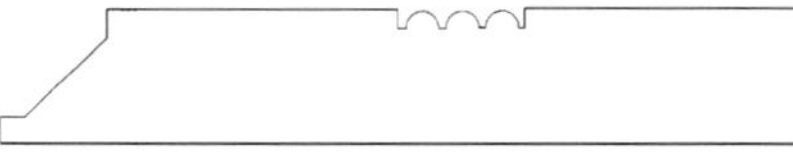

$^3/_4$ x 5$^1/_2$

Most Eastlake moldings have a bead or reed detail like this molding. Note it is not symmetrical and could be used with a back band if so desired.

HHCS012
Eastlake Casing

1890's
Victorian

$^3/_4$ x 5$^1/_2$

Another "Eastlake" casing with the familiar bead detail. This molding is actually very similar to HHCS011, but would carry the paint or shadow lines a little bit differently.

HHCS013
Queen Anne Casing
1890's
Victorian

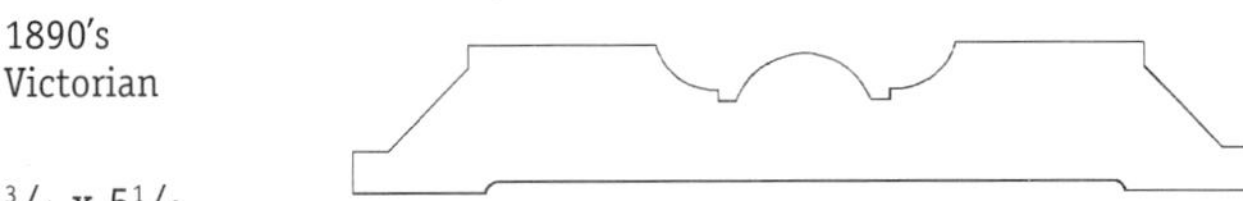

$^3/_4$ x $5^1/_2$

This casing was actually labeled in the catalog, which was not common for casings of this period. Meant for Queen Anne style houses, this molding has a center bead that is quite handsome. It is only seen in the 1890 catalog but it is a very distinctive style that would work on historic and modern Victorians.

HHCS014
Casing
1890's
Victorian

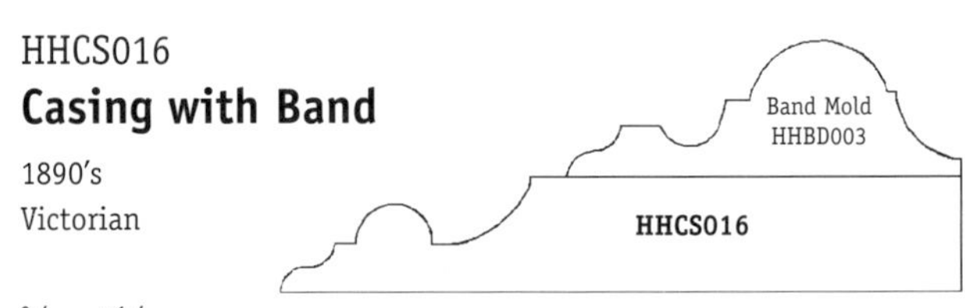

$^3/_4$ x $5^1/_2$

A simple Victorian casing that was shown with a band molding. This was a typical arrangement that is subdued by some Victorian standards.

HHCS015
Casing with Band
1890's
Victorian

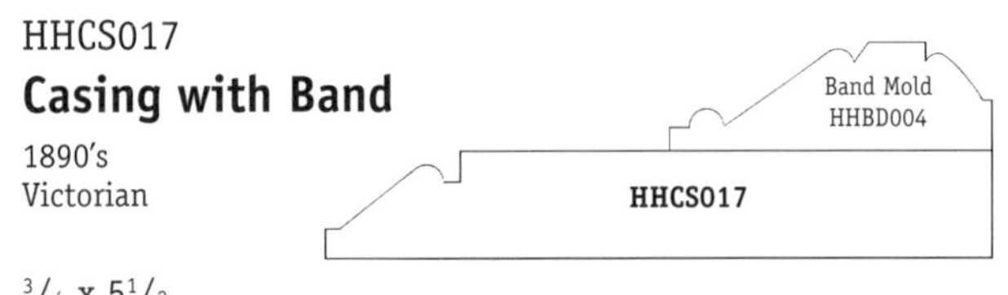

$^3/_4$ x $5^1/_2$

A simple casing as seen in the first standard molding catalog. The thumbnail or ovollo on the front edge is simple, but built-up with the band mold the overall appearance is quite attractive. The cove on the back edge is another interesting detail that increases the shadowing and charm.

HHCS016
Casing with Band
1890's
Victorian

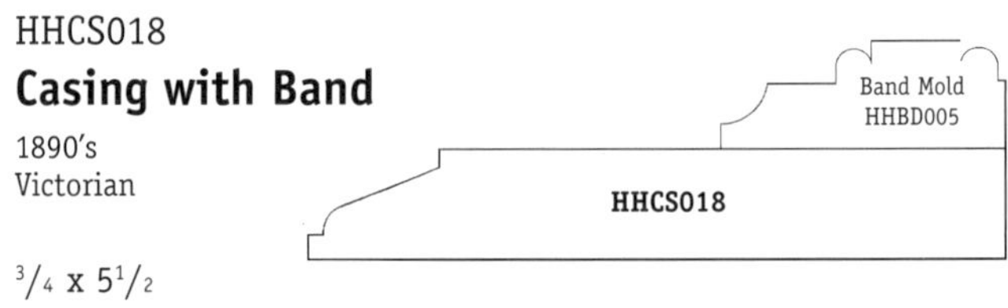

$^3/_4$ x $5^1/_2$

A Victorian molding with detailing we see in many panel molds and band molds. Building up moldings was a common Victorian detail and the band molding on this casing really creates a beautiful profile.

HHCS017
Casing with Band
1890's
Victorian

$^3/_4$ x $5^1/_2$

A handsome Victorian casing, especially when used in conjunction with the band mold shown. Simple and clean, it still has nice lines that would look great painted or stained.

HHCS018
Casing with Band
1890's
Victorian

$^3/_4$ x $5^1/_2$

Another simple Victorian casing that is only seen in the first standard catalog. The band molding shown is handsome and unique.

HHCS019
Casing with Band
1890's
Victorian

$^3/_4$ x $5^1/_2$

A Victorian casing similar to HHCS002. It was intended to accept a band mold like the one shown. It did not make it into the 2nd catalog edition.

HHCS020
Casing
1890-1915
Victorian

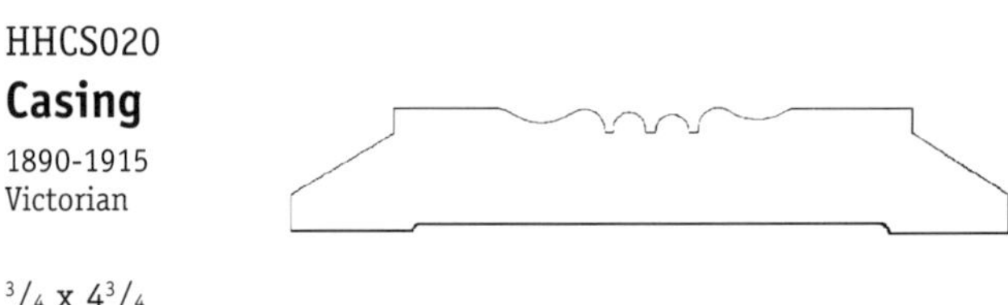

$^3/_4$ x $4^3/_4$

This Victorian casing has a longer life than most casings of this period. Symmetrical with the familiar bead detail, this casing was used most commonly with rosettes and headers.

HHCS021
Casing

1890-1915
Late Victorian

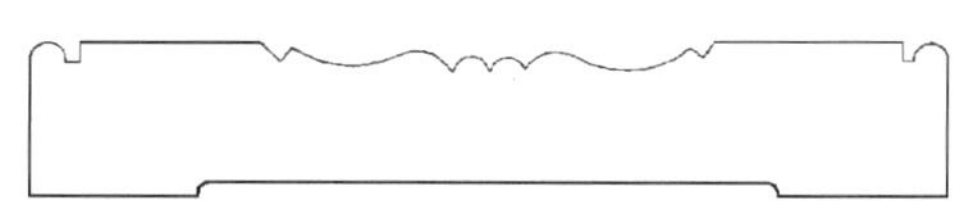

$^3/_4$ x $5^1/_4$

This Victorian molding has much simpler lines and is not nearly as bold as other casings of this period. It may be one reason why it lasted through to 1915. Still a handsome mold it would look great in a new home today seeking an authentic turn of the century look.

HHCS022
Casing

1890-1915
Late Victorian

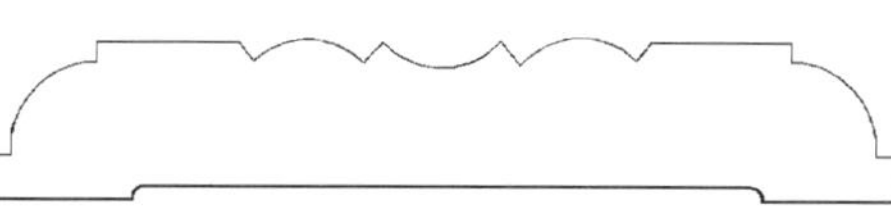

$^3/_4$ x $5^1/_4$

A symmetrical Victorian mold which also lasts through 1915. This molding is one of the more popular Victorian moldings and was also seen in Colonial Revival as well as Beaux Arts style homes at the turn of the century.

HHCS023
Casing

1890-1915
Late Victorian

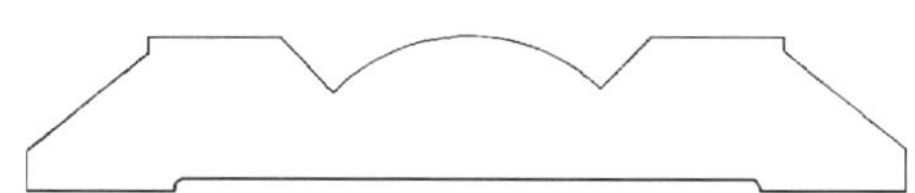

$^3/_4$ x $5^1/_4$

This molding has bold cuts that fit well into late Victorian as well as some early Arts & Crafts style homes.

HHCS024
Casing

1890-1915
Late Victorian

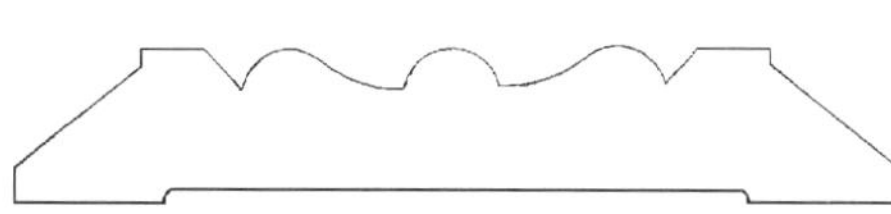

$^3/_4$ x $5^1/_4$

A popular Victorian casing motif is the center bead or reed that runs down the middle of these moldings. By establishing a center line it highlights the symmetrical nature of the casing and would also line out (i.e. define) windows and doors, especially when painted.

HHCS025
Casing

1890-1940
Classical

$^3/_4$ x 4 or 6

This classical molding appears in standard catalogs into the 1940's. The bead and cove profile was used in Victorian times with a band mold, and with a back mold or by itself in the Arts & Crafts, Period Revival, and finally the Colonial style in the 1940's. It did change in size over time, after the 1920's, shrinking from 6 inches to 4 inches. Be sure to order it in the correct size, depending on the period design you need.

HHCS026
Casing

1890-1915
Late Victorian

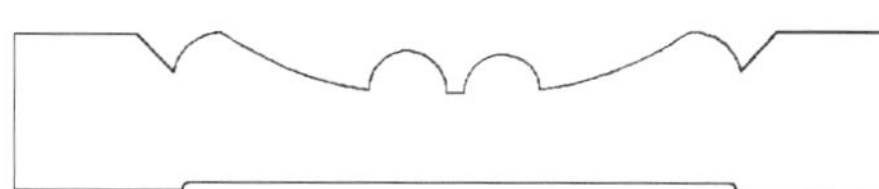

$^3/_4$ x $5^1/_4$

A typical Victorian mold with dual center beads that establish the center line. This molding lasts into the Arts & Crafts period probably because of the square edges which complement that style.

HHCS027
Casing

1900-1926
Classical

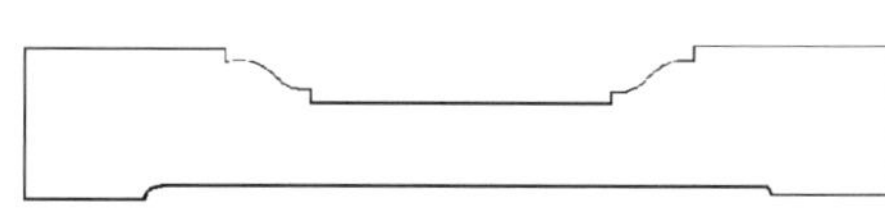

$^3/_4$ x $4^3/_4$

This molding was newly introduced in the 2nd catalog edition. It is symmetrical with a raised OG detail, but the square outside edge speaks more of simple and classical Arts & Crafts styling. This classic casing sees wide use up through the 1920's.

HHCS028
Casing

1900-1915
Late Victorian

$^3/_4$ x $5^1/_4$

A late Victorian casing, it has square edges like some Arts & Crafts style moldings, but the center portion is all Victorian. Similar to other moldings of this period. it is symmetrical and would work well with rosettes and header blocks.

HHCS029
Casing
1900-1915
Late Victorian

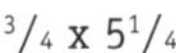

$^3/_4$ x 5$^1/_4$

This late Victorian casing is typical of the change in architectural styles from the rich Victorian to the simple Arts & Crafts. The lines are much more subdued than what we see before. It is still symmetrical, but reflecting the decline in usage of rosettes and headers after 1900, could be mitred at the corner as well.

HHCS030
Casing
1900-1915
Late Victorian

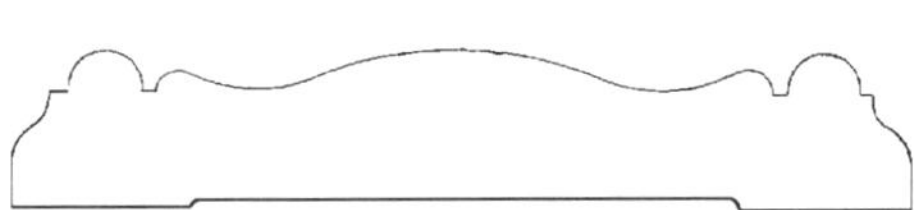

$^3/_4$ x 5$^1/_4$

A late Victorian mold that is a little more Victorian than Arts & Crafts, yet is still transitional. The bead detail on the outside edges helps maintain its symmetrical shape and keeps this molding looking great for today's homes as well.

HHCS031
Casing
1900-1915
Late Victorian

$^3/_4$ x 5$^1/_4$

A symmetrical Victorian casing with a bar that establishes the center line of the molding. This detail helps create the effect of framing out the door and window.

HHCS032
Casing
1900-1925
Late Victorian

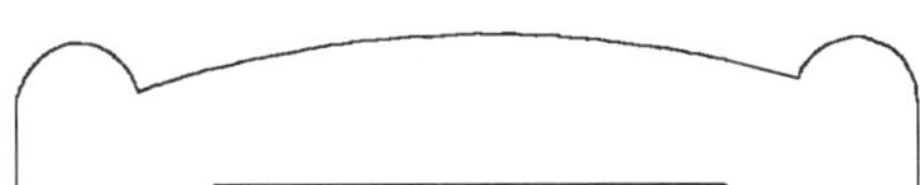

$^3/_4$ x 5$^1/_4$

This was a very popular late Victorian mold that was simple and clean like the Arts & Crafts style, yet the bead helped it work with many of the transitional designs. Used most often with rosettes and headers, it was popular and important to this period.

HHCS033
Casing
1900-1915
Late Victorian

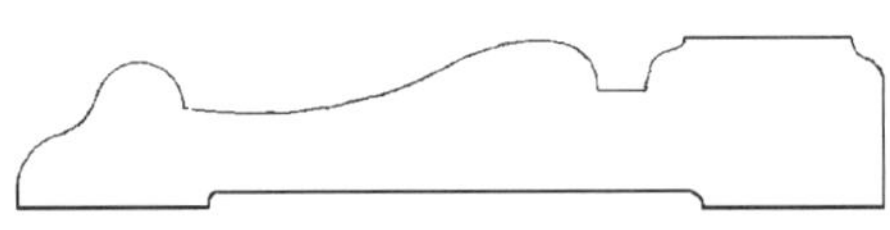

$^3/_4$ x 4$^1/_4$

A late Victorian mold which breaks with the symmetrical designs we see so often. This molding has the typical bold Victorian lines but reflects its later introduction in that it was not meant to be used with rossettes and headers.

HHCS034
Casing
1900-1915
Late Victorian

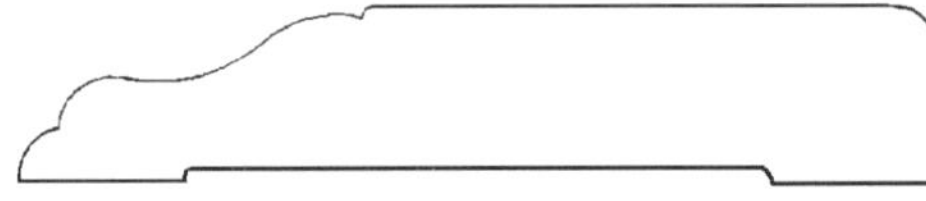

$^3/_4$ x 4$^1/_4$

A simpler version of HHCS035, this casing could be used with a back band to build it up if so desired. This casing evokes an Arts & Crafts simplicity which allows it to work on many new homes quite gracefully.

HHCS035
Casing
1900-1915
Late Victorian

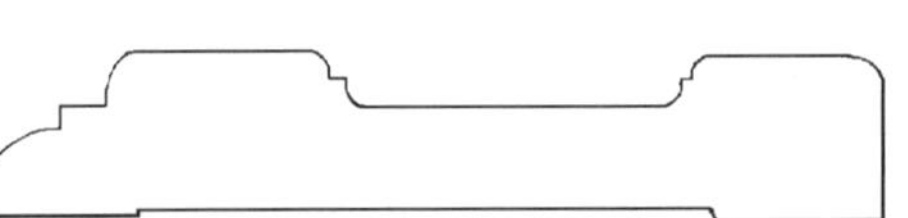

$^3/_4$ x 4$^3/_4$

A kind of lopsided symmetrical casing. Obviously in retrospect it is an effort to combine symmetrical and asymmetrical design into one look. Simpler detailing is typical of this transitional period and would be a very distinctive casing if used today.

HHCS036
Casing
1900-1915
Late Victorian

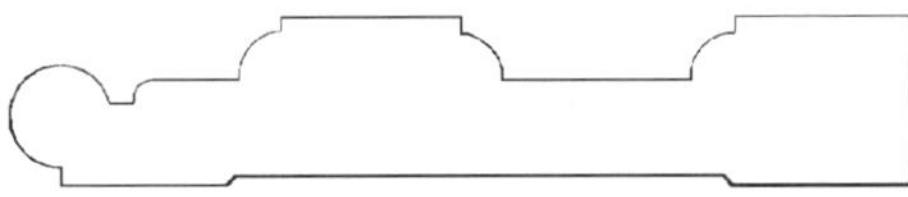

$^3/_4$ x 4$^3/_4$

A very bold late Victorian casing which is really a hold over from earlier Victorian styles. Asymmetrical in design, the bead detail is a great effect that strongly defines the outside edge of the door or window.

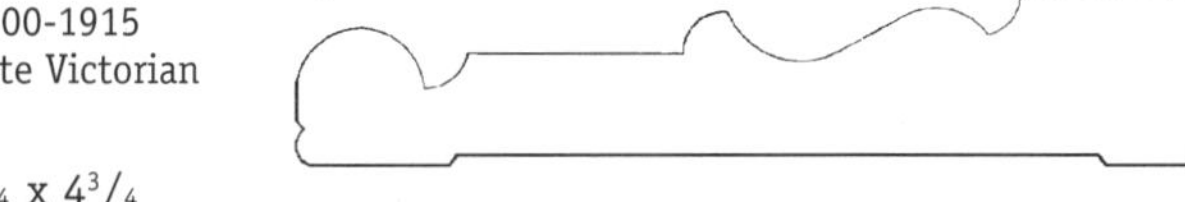

HHCS037
Casing
1900-1915
Late Victorian

$^3/_4$ x $4^3/_4$

Another bold late Victorian mold that is much more Victorian than transitional. There are some classical influences in this mold with the bead on the front edge.

HHCS038
Casing
1900-1915
Late Victorian

$^3/_4$ x $4^3/_4$

This casing is a little more transitional in that it has simpler details and the lines are not as bold. This molding would look great with a back band and has some classical detailing that makes it quite contemporary.

HHCS039
Casing
1900-1915
Arts & Crafts
1915-1945 available with square edges

$^3/_4$ x $5^1/_4$

Here is an Arts & Crafts classic that was shown with round edges up to 1915 and with square edges up into the 1940's. It was even available with one edged rounded and the back edge square (we can also provide this version if you let us know when ordering). A simple yet versatile casing that can be dressed up with a back band or simply stained for a true Arts & Crafts feel.

HHCS040
Casing & Back Band
1900-1915
Late Victorian

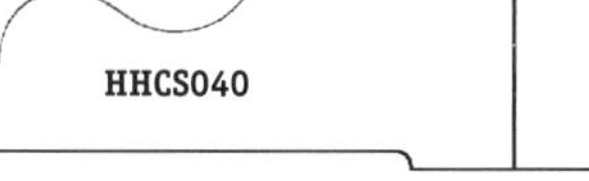

$^3/_4$ x $4^1/_4$

A late Victorian casing that is transitional. Some elements are bold as in the Victorian, but the rest is clean and subtle.

HHCS041
Casing & Back Band
1900-1915
Classical revival

$^3/_4$ x $4^1/_4$

A great casing that has some wonderful classical elements. The bead and cove on the face segues appealing to a long fillet and OG curve. The optional back band really compliments this casing and makes it a natural for many high style homes.

HHCS042
Casing & Back Band
1900-1915
Classical Revival

$^3/_4$ x $4^1/_4$

A transitional Victorian that has clean lines and bold details. Complimented appropriately by an optional back band, this casing is really unique and wonderful.

HHCS043
Casing & Back Band
1900-1915
Classical Revival

$^3/_4$ x $3^3/_4$

Originally going back to the 1870's and the days when casing were called architraves, this transitional casing echoes the beauty and balance of classical architecture. An appropriate fit for historic homes of this period, and can also work well in today's homes.

HHCS044
Casing & Back Band
1900-1920
Arts & Crafts

$^3/_4$ x $3^3/_4$

A very simple Arts & Crafts casing that sweeps back with clean steps. The optional back band compliments the round edges on the casing with a simple cove detail. This molding is available through the Arts & Crafts period and was quite popular.

HHCS045

Casing

1900-1915
Late Victorian

$^3/_4$ x $5^1/_4$

The last casing in the late Victorian catalog. It is a transitional molding that has some classical elements with the OG and over-sized bead and cove which would work well in today's homes.

HHCS046

Casing

1915-1920
Arts & Crafts

$^3/_4$ x $4^3/_4$

The moldings we find after 1915 are simple and clean in keeping with the Arts & Crafts tradition. This simple, symmetrical casing was handsome when mitered at the corners. Usually stained in the period, it can also work in historical and modern homes when painted.

HHCS047

Casing

1920s
Period Revival

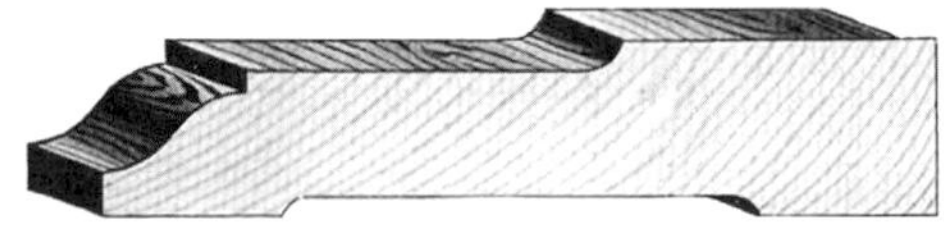

$^3/_4$ x $3^5/_8$

This casing shares many design characteristics with HHCS048. This is quite common in this period as slight design differences are made to create a different look. HHCS047 is the more classically inspired of the two and would work well with a back band of similar shape.

HHCS048

Casing

1920-1935
Period Revival

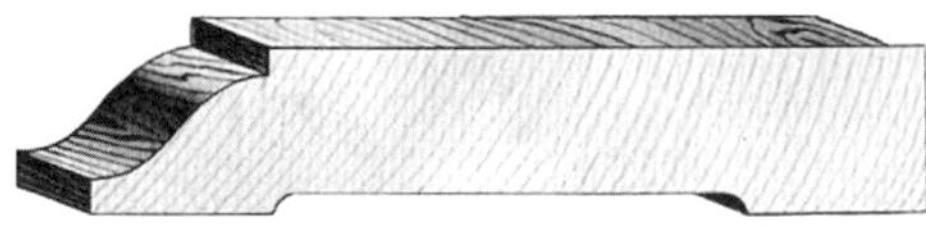

$^3/_4$ x $3^5/_8$

Like HHCS047, this molding's design represents a clean break from the Victorian looks which preceded it, yet there is enough design interest to allow this casing to enjoy popularity beyond the Period Revival era. It can even be paired with a back band to give it a bold and more classical appearance.

HHCS049

Casing

1930-1945
Period Revival

$^3/_4$ x $2^1/_4$

From 1930 into the 40's, many firms developed their own patterns of moldings that were not standard. This one seems to find a place in many of these post-1930 catalogs. Simple and bold, it looks great in homes of this period. Note the width is much thinner than earlier casings, as is typical of this era.

HHCS050

Casing

1935-1940
Colonial Revival

$^3/_4$ x $3^1/_4$

A classically inspired molding that is shrunk down for this period. In colonial times this casing might be 6 inches wide, in the 30's and 40's most casings were thinner like this one, which is offered at $3^1/_4$" wide. A great looking molding for historic or new homes.

HHCS051

Casing

1930-1945
Period Revival

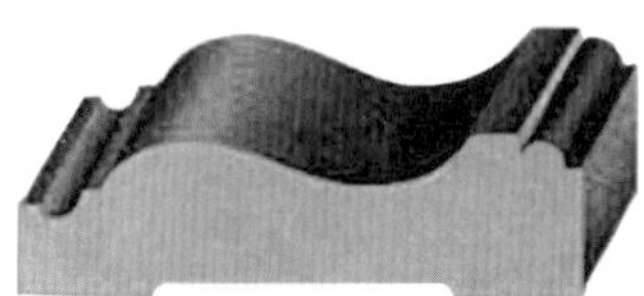

$^3/_4$ x $2^3/_4$

Another compact yet handsome casing of the 30's and 40's. This casing looks great when painted and can work in many style homes. Use this molding when the scale of the room calls for a thinner molding but boldness is still required.

HHCS052

Casing

1935-1940
Colonial Revival

$^3/_4$ x $3^1/_4$

In the 1940's, two of the most popular moldings were the beaded and reeded casings like HHCS052 and HHCS053. These two classics can take us back to Grandma's house and still look great today.

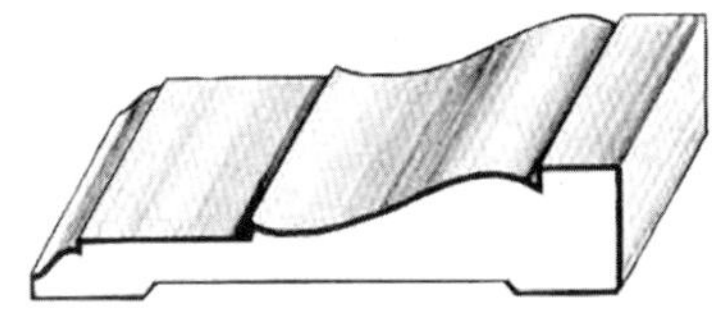

HHCS053
Casing

1940's
Colonial Revival

$^3/_4$ x $3^1/_4$

This pattern is in my house, and we are proud to reproduce it here for your remodel or historically inspired home.

HHCS054
Casing

1940's
Colonial Revival

$^3/_4$ x $3^1/_2$

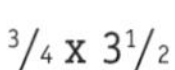

Another classically inspired molding that was popular in the 40's. Though not as wide as true original moldings, these are still unique and wonderful.

Band Moldings

Band moldings are a uniquely Victorian type of molding which were added to door and window casings to create additional design detail and complexity. Design books from the 1870's and 1880's show many casings (or "architraves" as they were called then) built up with two or even three different band and accent molds. As high Victorian style ebbs in popularity, band molds disappear. They are included in the first standard molding catalog in 1890, but not in any of the subsequent editions.

Most of these band molds resemble the shapes of panel molds and base caps. You will note that most band molds sit on top of a simple casing. You can use the Hull Historical Molding Catalog to select from our band molds and casings to develop your own custom molding package. Or if you would like Hull Historical to consult on your project, we can design complete packages for your new or historic home.

The band molding is replaced in later eras by the back band. The primary difference between the band mold and the back band is that band moldings are more ornate and sit on top of the casing, while back bands have simpler designs and sit on the back edge of the casing. Please see the Back Band section if you are looking to add detail and style to moldings from later eras such as Arts & Crafts or Period Revival.

HHBD001
Band Molding

1890's
Victorian

$^7/_8$ x $2^1/_2$

A great looking Victorian band mold which is built up and strongly featured. Note how the appearance of the simple casing is transformed by pairing it with this band mold.

HHBD002
Band Molding

1890's
Victorian

$^7/_8$ x $2^1/_2$

Very similar to HHBD001, but with front and rear edges that compliment the lower casing. The bolection or astragal is very prominent and a typical Victorian touch.

HHBD003
Band Molding

1890's
Victorian

$^7/_8$ x $2^1/_2$

A heavy bolection band mold with a rear edge that well suited for the casing. This detail is very popular in this period and is an excellent design addition to Victorian doors and windows.

HHBD004
Band Molding

1890's
Victorian

$^7/_8$ x $2^3/_4$

This Victorian molding that looks great in both historic and new homes. The bead detail on the front edge is charming, and when painted plays with the light wonderfully.

HHBD005
Band Molding

1890's
Victorian

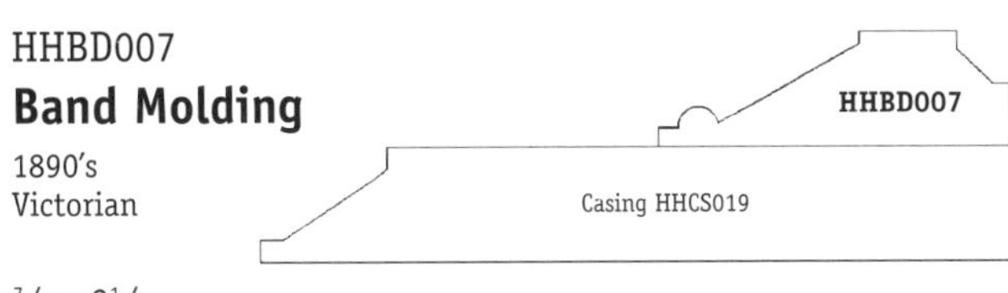

$^7/_8$ x $2^1/_4$

Another appealing Victorian band mold that is bold and interesting. The bead detail is a popular one and works well on the outside edge of this casing.

HHBD006
Band Molding

1890's
Victorian

$^7/_8$ x $2^1/_4$

This band mold is similar to HHBD007. The slanted face compliments the casing, while the bead is used to add charm.

HHBD007
Band Molding

1890's
Victorian

$^7/_8$ x $2^1/_2$

This band mold is similar to HHBD006. The slanted face mirrors the casing design.

HHBD008
Band Molding

1890's
Victorian

$1^1/_8$ x $3^3/_4$

This band is bigger than most. The mold features two bolections or astragals to create a bold appearance. The OG face would look great with casing HHCS014.

HHBD009
Band Molding

1890s
Victorian

$1^3/_8$ x $4^1/_4$

This is our biggest band mold (almost big enough to be used by itself!). It was also very popular in its day and you will note it is very similar to HHBD008, except for the back edge which is a little more ornate with the classic bead and cove profile.

HHBD010
Band Molding

1890's
Victorian

$^7/_8$ x $2^3/_4$

An appealing yet simple band mold that is not as bold as the bolection or reed-type band mold, yet still has nice lines that look great when painted. The back edge has an ovollo detail that is quite handsome.

HHBD011
Band Molding

1890s
Victorian

$^7/_8$ x $2^3/_4$

This band is similar to HHBD012, except the back edge has a bead and cove detail as opposed to the ovollo. This band mold is perfect in pre-1900 Victorians.

Back Band Moldings

Back Bands first appear in the catalogs in 1900 and continue to be offered into the 1930's. They are most popular in the 1910's and 20's and are used in many Period Revival homes of that time. The back band replaces the band mold in the catalogs and, as discussed in the band mold section, is really just a more subtle version of this mold.

The back band gives casings a more traditional look, harkening back to the days when a casing was called an architrave (after the name in classical architecture). Just as the outside edge of the architrave was built up and often molded, the back band adorns the casing to add depth and interest.

I think back bands are a great way to spice up moldings today. It adds a classic feel to any normal casing and gives life to a great molding package.

HHBB001
Back Band

1900-1940
Classical

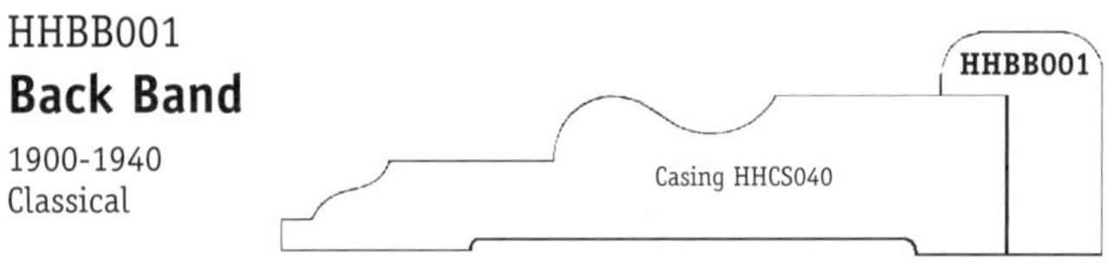

$1^1/_8$ x $1^1/_8$

This simple back band lasts the longest in the catalogs because it provides a classic built-up feel suitable for any period. The clean lines and eased edges are unobtrusive yet bold. This is an easy way to make any casing special.

HHBB002
Back Band

1900-1915
Classical Revival

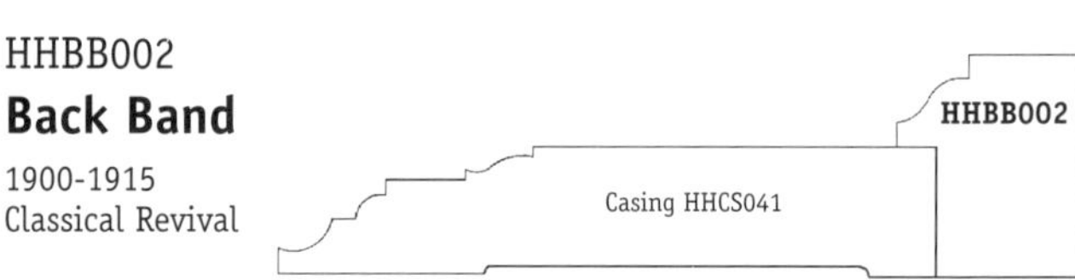

$1^1/_8$ x $1^3/_8$

Tall, bold classic, the cyma recta detail on this back band is right out of Palladio's sketch book. The strong sense of classicism is seen as the back band harmoniously interplays with the casing to create a great look.

HHBB003
Back Band

1900-1915
Classical Revival

$1^1/_8$ x $1^3/_8$

This back band gets its classic feel from the timeless OG profile. The back band works well with the casing shown, but it would work well with many others also.

HHBB004
Back Band

1900-1920
Classical Revival

$1^3/_8$ x $1^3/_8$

The bead and cove profile on this back band goes well with doors and other millwork using the classic bead and cove profile. Note the front edge of the casing also uses the bead and cove detail for continuity.

HHBB005
Back Band

1900-1920
Classical Revival

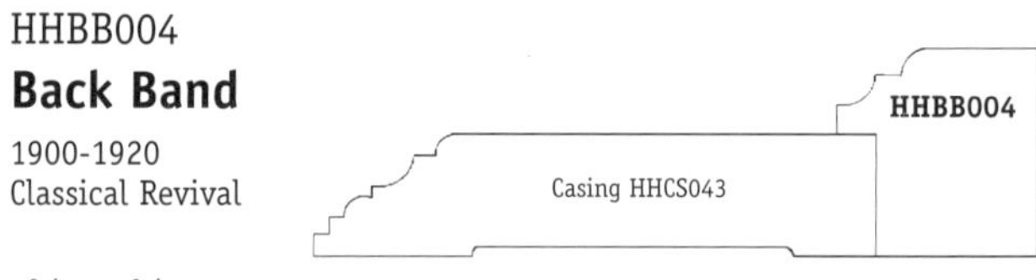

$1^1/_8$ x $1^1/_8$

This simple back band was very popular and we see it time and time again on the houses of this period. It looks great painted and creates a nice shadow line when combined with the right casing. I like it with the casing shown, or it can work on a flat plain casing.

Stops

Stops are a versatile and simple detail that add visual allure to any millwork. Used most commonly with doors or windows, (stops usually are fastened to the jamb), the name comes from the function of the piece, to stop movement or to hold a detail (especially movable elements) in place. Stops were used everywhere, from holding glass in a sash or frame to keeping panel frames from moving.

The variety of stops shown here reflect the wide variety of purposes for which they were used. For this reason stops were generally offered in many sizes, varying not only in width but thickness as well. We offer two thicknesses of the early profiles, $3/8''$ and $1/2''$, just as in the Victorian catalogs. Sometimes they were offered in very wide thicknesses for very large jam openings. The $1/2''$ is most commonly used today but the $3/8''$ may be what you need to properly match a charming historic detail in your home. Note the last two stops were thicker, reflecting changing styles in the 30's.

HHST001
Stop
1870-1890
Victorian

$3/8$ x $1^1/2$

$1/2$ x $1^1/2$

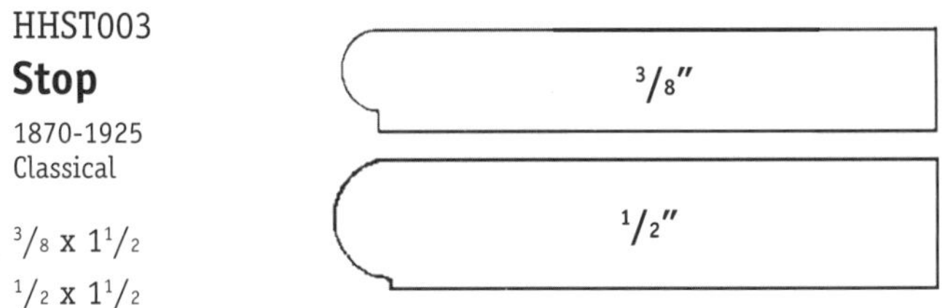

This is a great looking stop, probably my favorite. We first see it in an 1870's catalog and in the first standard catalog of 1890. The bead that moves up to an OG is a small detail but one that shows great lines. The fillet below the bead may seem tiny but it creates a great shadow line that will show well in any room.

HHST002
Stop
1870-1945
Classical

$3/8$ x $1^1/2$

$1/2$ x $1^1/2$

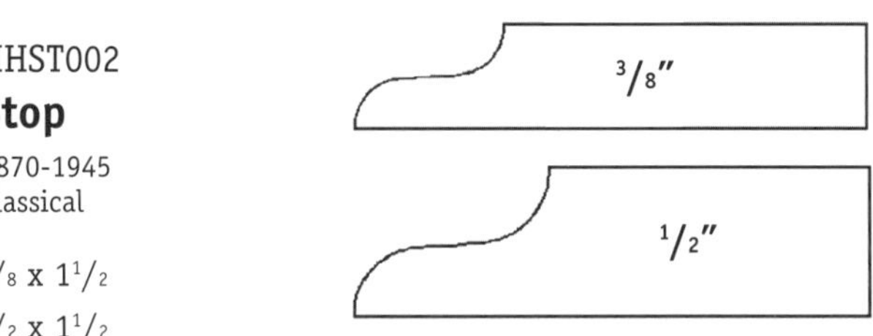

The classic OG stop can still be seen in some lumber yards today but our OG, the exact profile which was offered in historic catalogs, is much more bold and deep than the modern alternatives. Available in $3/8$ inch and $1/2$ inch thickness, it is a timeless detail that looks great on any historic or new home.

HHST003
Stop
1870-1925
Classical

$3/8$ x $1^1/2$

$1/2$ x $1^1/2$

This stop was offered in the catalogs for over fifty years. The bead detail is typical of Victorian moldings, but was versatile enough to be used widely with Period Revival homes also. We often incorporate this stop on our historical projects.

HHST004
Stop
1900-1945
Classical

$1/2$ x $1^1/2$

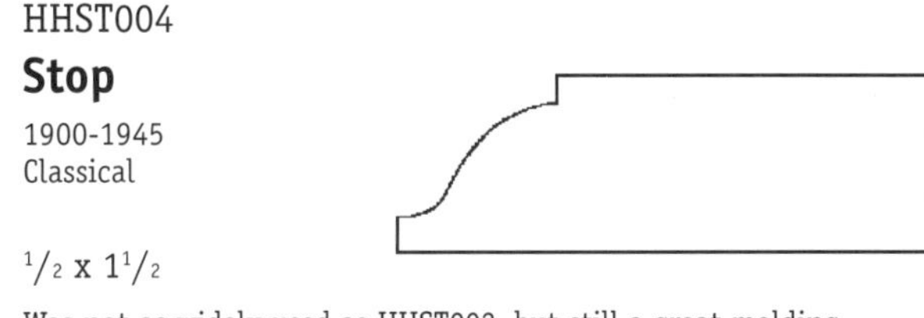

Was not as widely used as HHST002, but still a great molding. Simple in detail, it can work on new homes because its classical styling complements many decors.

HHST005
Stop

1890-1915
Victorian

$^3/_8$ x $1^1/_2$
$^1/_2$ x $1^1/_2$

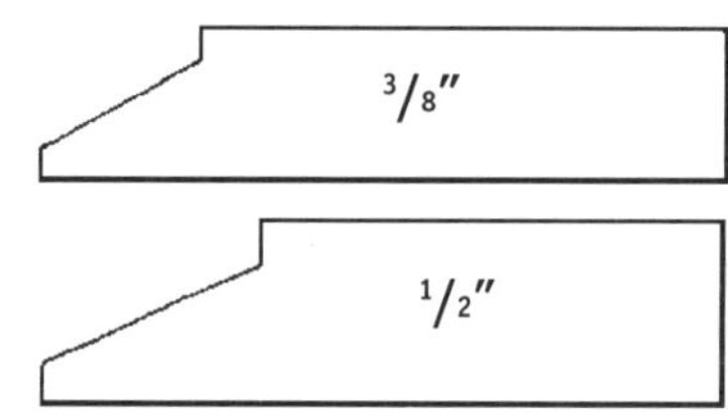

This is a very Victorian stop. Similar to a PG profile on doors, the sharp angle used on this stop was popular in some panel molds as well.

HHST006
Stop

1920-1945
Arts & Crafts

$^1/_2$ x $1^1/_2$

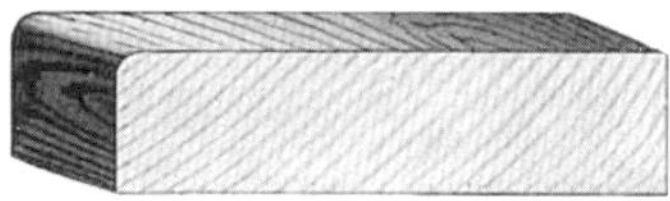

A true Arts & Crafts stop that, interestingly, doesn't show up in the standard catalog until 1920. It lasts into the 1940's catalogs and can still be found in some form today.

HHST007
Stop

1930-1945
Eclectic

$^3/_4$ x $1^1/_2$

The ovollo detail on the front edge is another element of the classical language of moldings. Despite its classical appearance, it borrows from many styles and fits well into the eclectic molding designs from the 1930's.

HHST008
Stop

1930-1945
Eclectic

$^3/_4$ x $1^1/_2$

This stop is Eclectic and offers a different take on the bead and cove profile of earlier eras. Much like the HHST007, its back edge is routed out to make a tighter fit or to allow for weather-stripping. A unique mold, it can make quite a statement in new or historic homes.

Window Stools

Window stools are an historic detail that are rarely used today. The window stool was used at the bottom of a window frame to create the window ledge.

The window stools from historic catalogs may appear similar at first glance but each one has distinctive details. Normally just the front edges are unique but these details can have a dramatic effect on the room. One thing that makes this collection of sills unique is their $1^1/_8$-inch thickness. Though it may not seem like much, the eye picks up the change from $3/_4$-inch trim to the thicker stool, serving to ground a window and give it a secure, solid feel.

In my opinion nothing finishes out a window like a thick window stool with a handsome apron. A classic look for new or historic homes.

HHWS001
Window Stool

1890-1900
Victorian

$1^1/_8$ x $3^1/_4$

Our earliest window stool picks up the beaded detail so popular in the Victorian era. The step down is reminiscent of many casings and band moldings and would go well with a slant face detail. It was only available for the first universal catalog and is rarely seen today.

HHWS002
Window Stool

1890-1945
Classical
with slight variations

$1^1/_8$ x $3^1/_4$

This classic window stool is still available today in some catalogs. Ours is different because it follows the proportions of the original. A full $1^1/_8$ inch thick, it has a bold appearance that really grounds the casing and window trim. This classic is just right for the historic home.

HHWS003
Window Stool

1890-1945
Classical

$1^1/_8$ x $4^1/_2$

A variation of HHWS002, it has more shape and thus more pronounced shadow lines. These two stools are popular through out our period of study. Owners of old homes will find this design very versatile for restoration projects. New home owners will find that it can work with many kinds of looks.

HHWS004
Window Stool

1900-1945
Arts & Crafts

$1^1/_8$ x $3^3/_4$

Simple and clean is sometimes hard to top. The full $1^1/_8$-inch stool here serves as a solid grounding line for windows in a room. Originally for the Arts & Crafts style, it lasted into the 40s'.

HHWS005
Window Stool

1900-1915
Late Victorian

$1^1/_8$ x $4^1/_4$

A unique and different stool, it was only seen in the 2nd edition of the catalog. Almost a transitional stool, it has some shape but is still plain. I recommend this stool primarily in painted applications; in my opinion it just does not look as good when stained.

HHWS006
Window Stool

1900-1915
Late Victorian

$1^1/_8$ x $4^1/_4$

This Victorian stool plays up the popular bead detail. This stool is perfect if you are looking to call attention to the detail on your windows. It is the full $1^1/_8$ inch thick and throws off great shadows and light whether painted or stained.

Cap Moldings

One of the joys in putting together this catalog is reintroducing great moldings like these. I was surprised to discover they are offered only briefly in the standard catalogs and, sadly, they have been almost completely lost today. The cap molding was popular after the turn of the century as a way to add a classical touch to the head of a door or window. The cap functions to give the header the look of an entablature and is one of the most cost-effective ways to add design interest to your doors and windows.

HHCM001
Cap Molding
1900-1925
Classical

$1^1/_8$ x $1^3/_8$

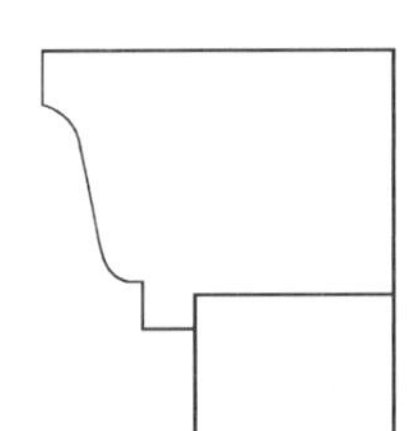

This cap mold uses the OG profile to top off a door or window. The upper detail resembles an entablature and works well on new or historic homes.

HHCM002
Cap Molding
1900-1925
Classical

$1^1/_8$ x $1^3/_4$

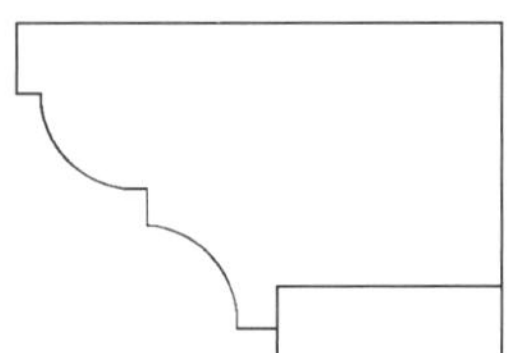

The classical cornice in an entablature should be at a 45-degree angle like this cap mold. The bead and cove detail is true and clean. A great header piece, it looks good painted or stained.

HHCM003
Cap Molding
1900-1915
Classical

$1^1/_8$ x $1^3/_4$

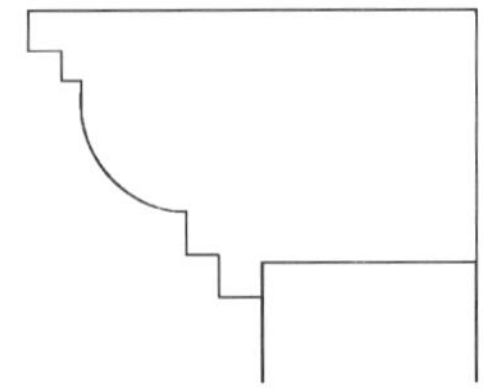

This bold cap mold is beautifully styled and detailed. The double fillets flanking the prominent bead exude classic molding design and yet is very flexible in application.

HHCM004
Cap Molding
1900-1925
Classical

$1^3/_8$ x $2^1/_2$

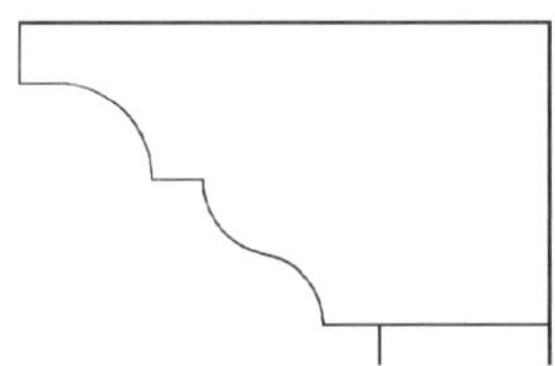

Like HHCM002, but this molding projects further out from the wall. It has many classical elements and was popular into the 20's.

HHCM005
Cap Molding
1900-1925
classical

$1^1/_8$ x $1^3/_8$

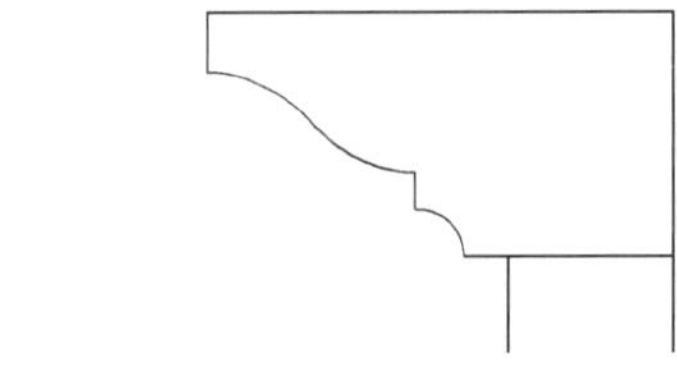

This is a simple yet interesting cap that really projects out for a nice hood effect. Another popular cap in its time, this OG and cove design works in many homes, historic or new.

HHCM006
Cap Molding
1900-1925
Classical

$1^3/_8$ x $2^1/_2$

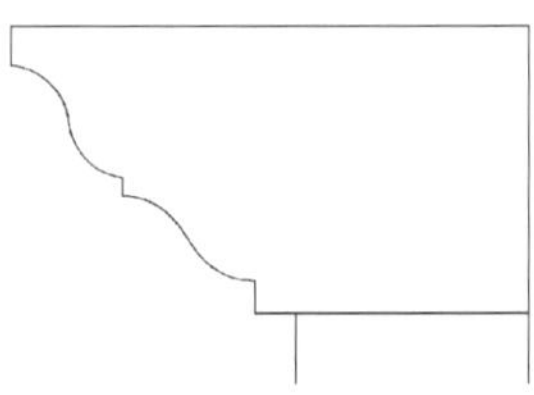

If one OG profile is good than two must be twice as good! Actually, the double OG look is very balanced and rhythmic. We recommend applying it with back bands HHBB002 or HHBB003.

Rosettes, Corner Blocks and Headers

This collection of Victorian header and corner blocks is unique and daring. Probably the single greatest challenge we faced in bringing these moldings back was duplicating the quality of the moldings at an affordable price. However, we know of no better way to bring authenticity to your Victorian project than with these types of moldings.

These were non-universal moldings; that means that they did not appear in the standard moldings section but appeared in the general body of each manufacturer's catalog. Highly decorative and unique, these were just too interesting to leave out.

The 1890 catalog did not recognize rosettes as a molding type. Instead rosettes, as they are called today, they were called "corner blocks." We have separated the rosettes from the corner blocks, though they were used for the same purpose. The rosettes feature circular designs while corner blocks are more eclectic. Corner blocks and rosettes were used not only at the inside corners of doors and windows but also as decorative blocks at wainscot level, or wherever the architect or builder might want to add decoration. (Feel free to intersperse them to meet the design needs of your home.) Header blocks are taller and appeared at the top corners of windows and doors.

As you may suspect, many of these are hand carved, so lead time may be greater on these products. All rosettes and header blocks are $1^3/_8$ inch thick.

HHRS001
Rosette
1885-1915
Victorian

$4^1/_2$ x $4^1/_2$
or $5^1/_2$ x $5^1/_2$

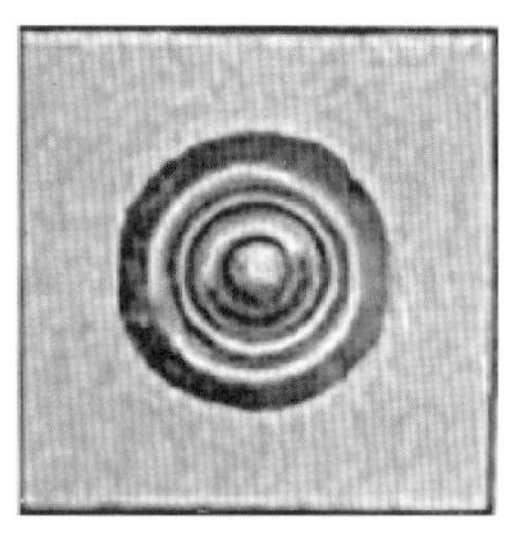

HHRS002
Rosette
1885-1915
Victorian

$4^1/_2$ x $4^1/_2$
or $5^1/_2$ x $5^1/_2$

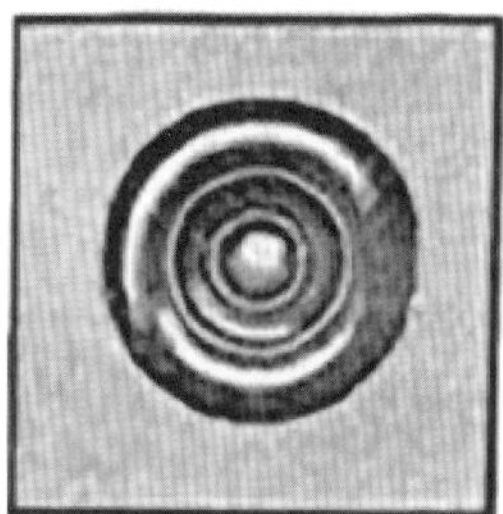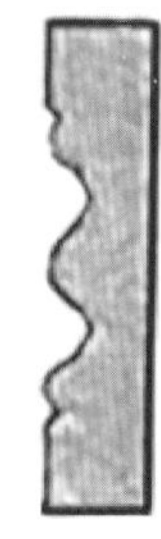

HHRS003
Rosette
1885-1915
Victorian

$4^1/_2$ x $4^1/_2$
or $5^1/_2$ x $5^1/_2$

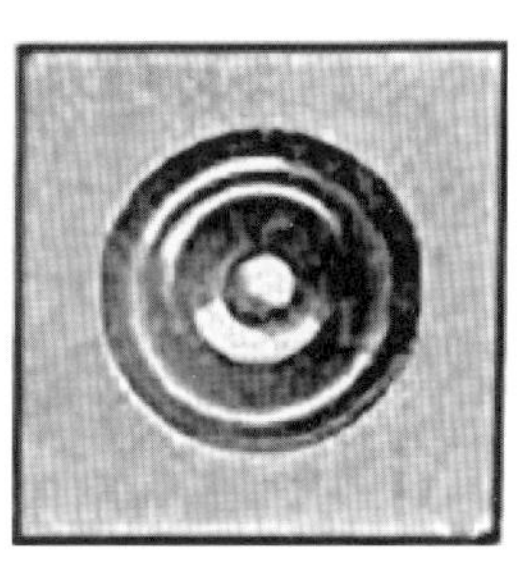

HHRS004
Rosette
1885-1915
Victorian

$4^1/_2$ x $4^1/_2$
or $5^1/_2$ x $5^1/_2$

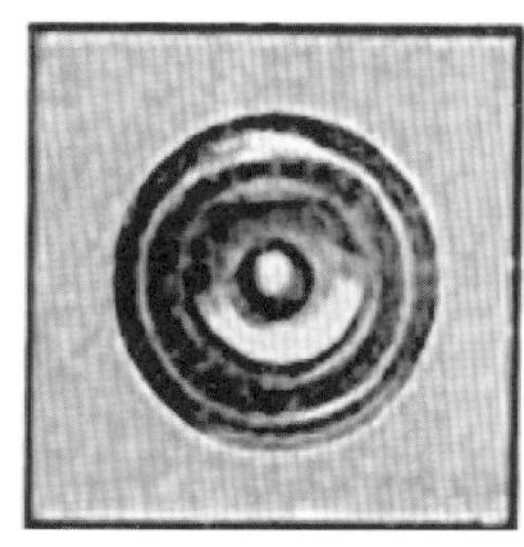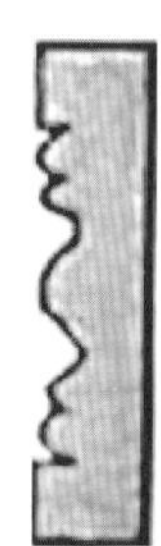

HHRS005
Rosette

1885-1915
Victorian

$4^1/_2$ x $4^1/_2$
or $5^1/_2$ x $5^1/_2$

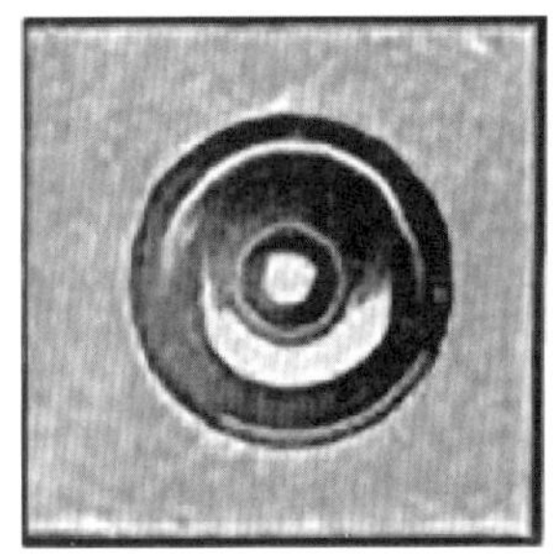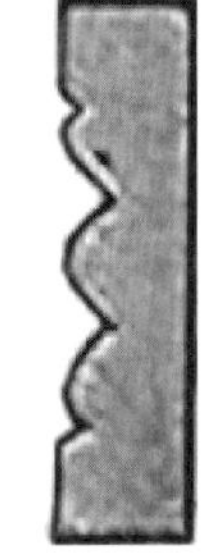

HHRS006
Rosette

1885-1915
Victorian

$4^1/_2$ x $4^1/_2$
or $5^1/_2$ x $5^1/_2$

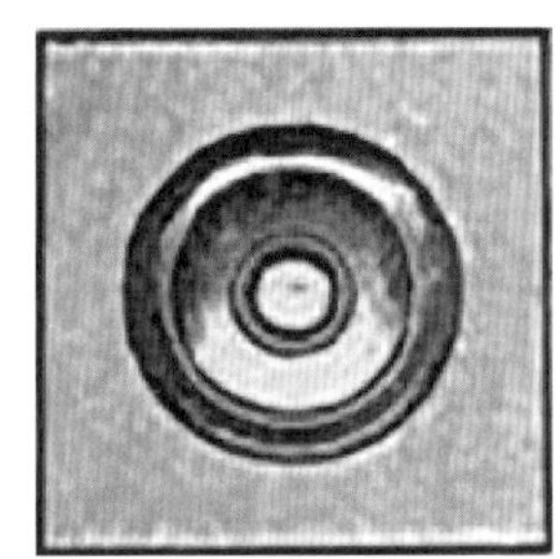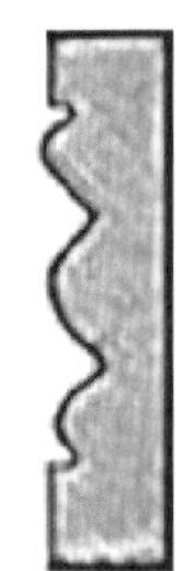

HHRS007
Rosette

1885-1915
Victorian

$4^1/_2$ x $4^1/_2$
or $5^1/_2$ x $5^1/_2$

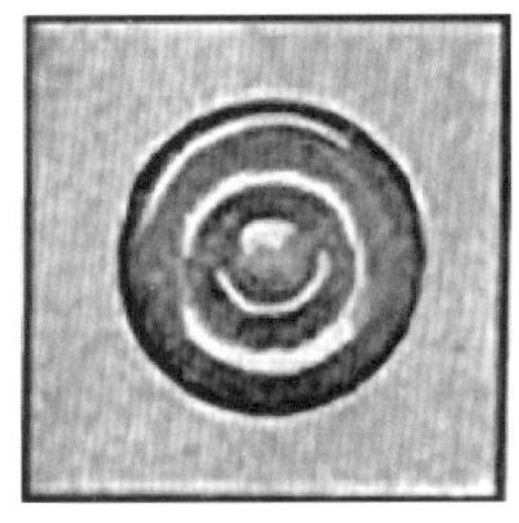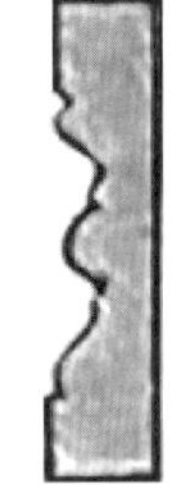

HHRS008
Rosette

1885-1915
Victorian

$4^1/_2$ x $4^1/_2$
or $5^1/_2$ x $5^1/_2$

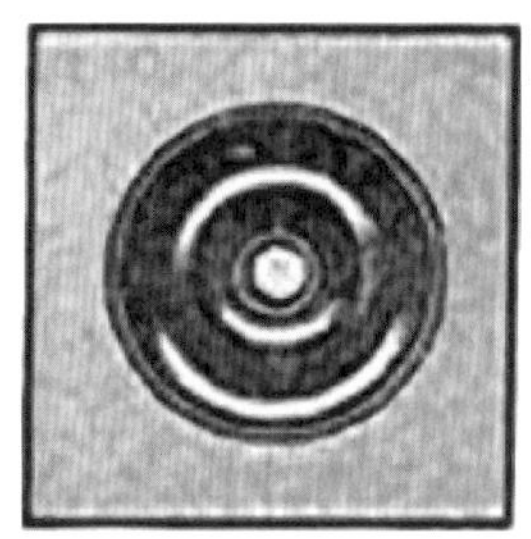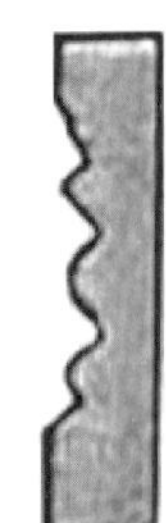

HHRS009
Rosette

1885-1915
Victorian

$4^1/_2$ x $4^1/_2$
or $5^1/_2$ x $5^1/_2$

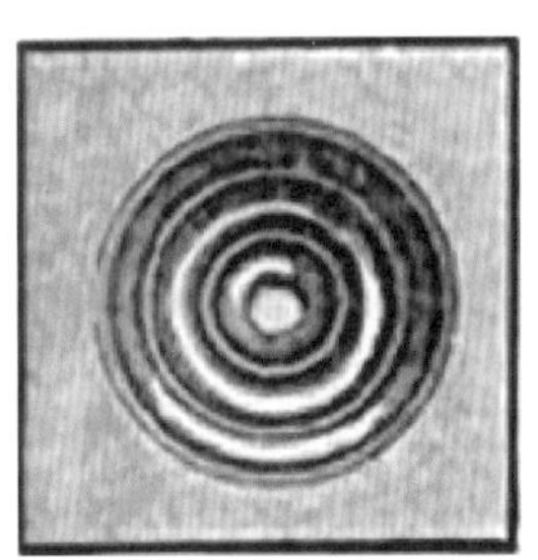

HHRS010
Rosette

1885-1915
Victorian

$4^1/_2$ x $4^1/_2$
or $5^1/_2$ x $5^1/_2$

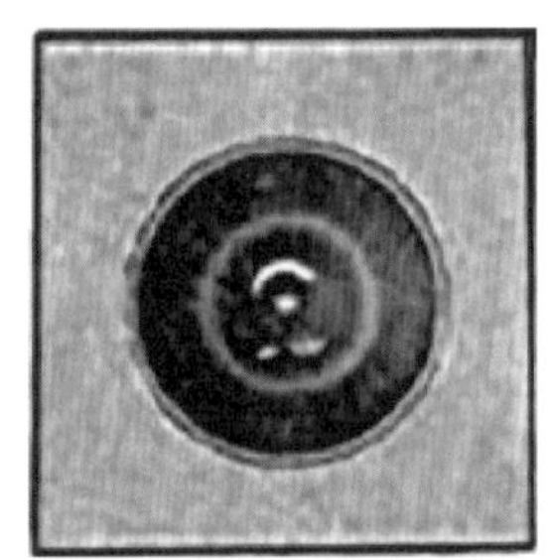

HHRS011
Rosette

1885-1915
Victorian

$4^1/_2$ x $4^1/_2$
or $5^1/_2$ x $5^1/_2$

HHRS012
Rosette

1885-1915
Victorian

$4^1/_2$ x $4^1/_2$
or $5^1/_2$ x $5^1/_2$

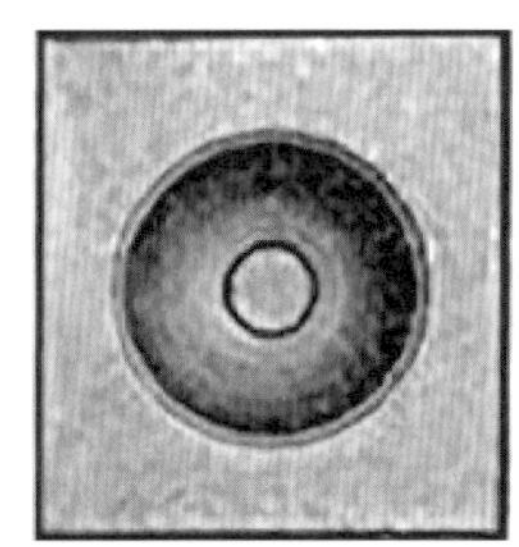

HHRS013
Rosette
1885-1915
Victorian

$4^1/_2$ x $4^1/_2$
or $5^1/_2$ x $5^1/_2$

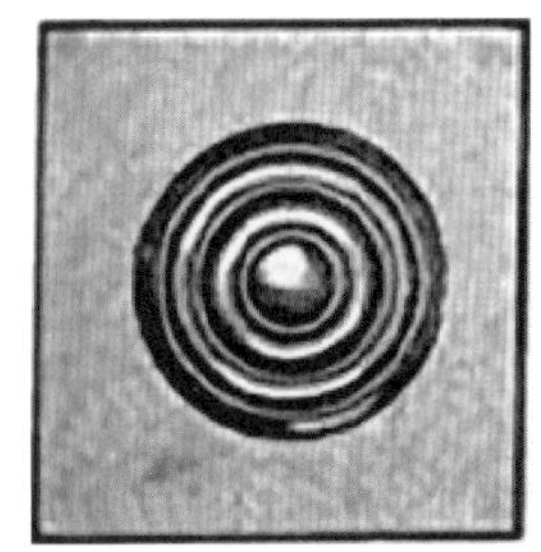

HHRS014
Rosette
1885-1915
Victorian

$4^1/_2$ x $4^1/_2$
or $5^1/_2$ x $5^1/_2$

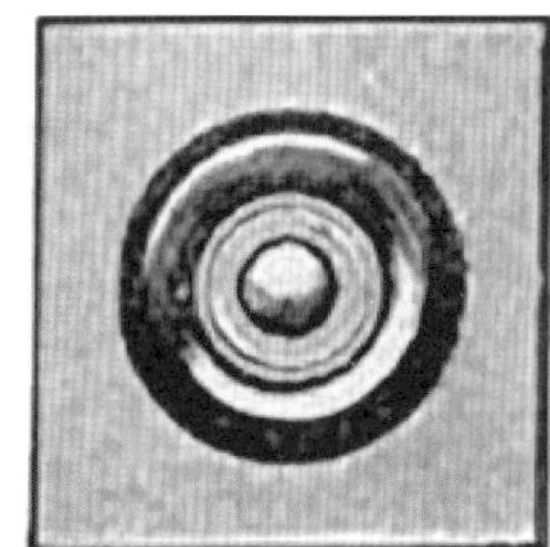

HHRS015
Rosette
1885-1915
Victorian

$4^1/_2$ x $4^1/_2$
or $5^1/_2$ x $5^1/_2$

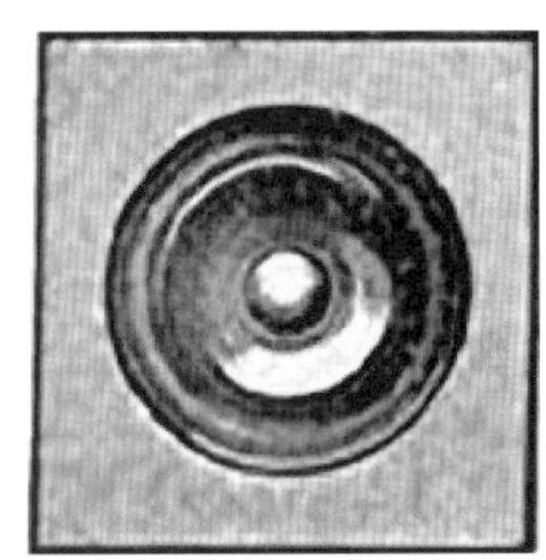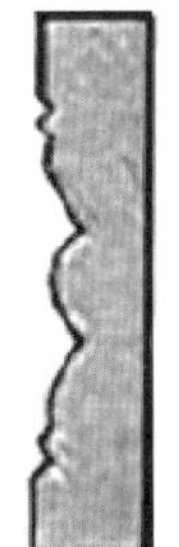

HHRS016
Rosette
1885-1915
Victorian

$4^1/_2$ x $4^1/_2$
or $5^1/_2$ x $5^1/_2$

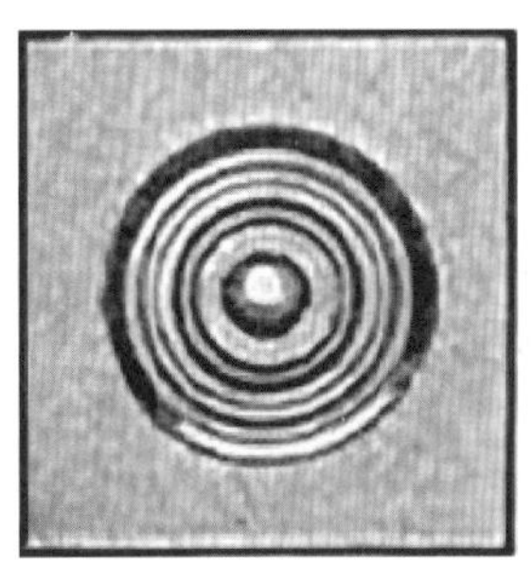

HHRS017
Rosette
1885-1915
Victorian

$4^1/_2$ x $4^1/_2$
or $5^1/_2$ x $5^1/_2$

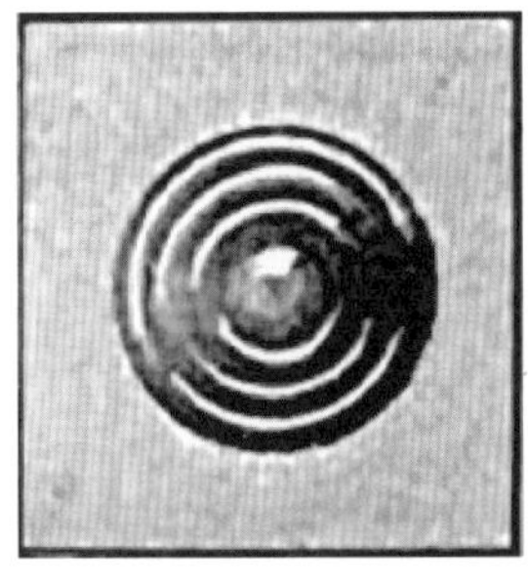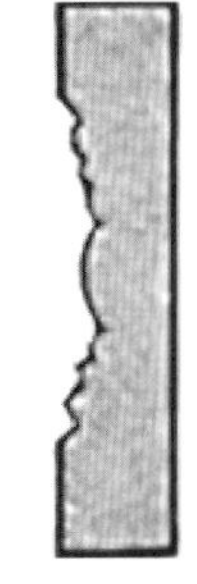

HHRS018
Rosette
1885-1915
Victorian

$4^1/_2$ x $4^1/_2$
or $5^1/_2$ x $5^1/_2$

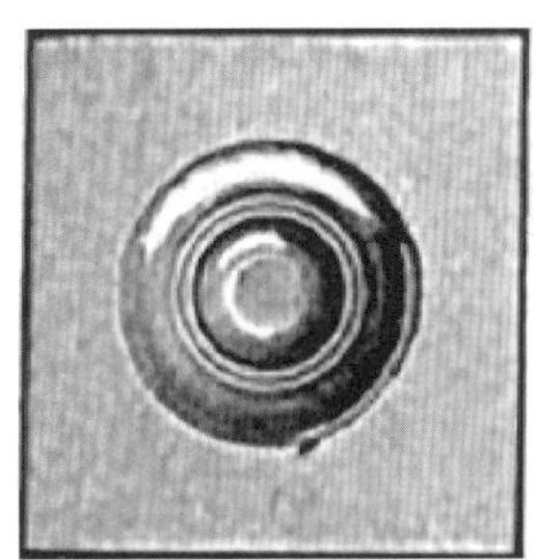

HHCB001
Corner Block
1885-1915
Victorian

$4^1/_2$ x $4^1/_2$
or $5^1/_2$ x $5^1/_2$

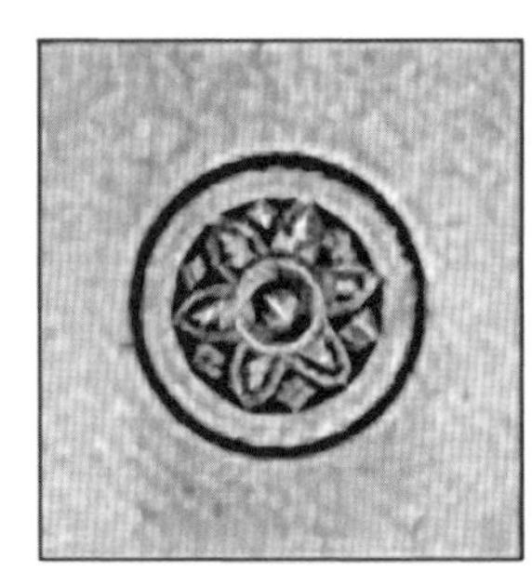

HHCB002
Corner Block
1885-1915
Victorian

$4^1/_2$ x $4^1/_2$
or $5^1/_2$ x $5^1/_2$

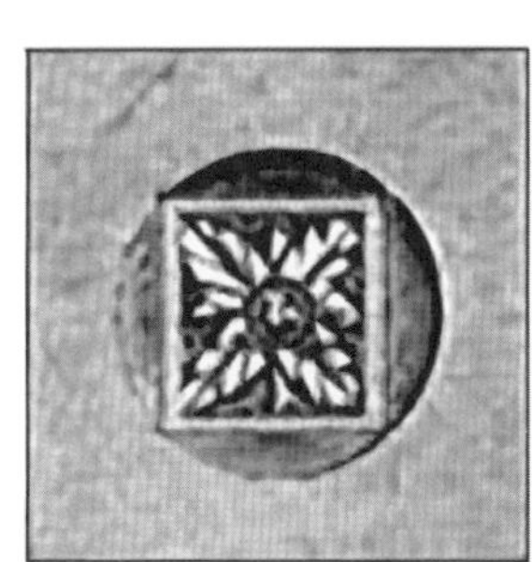

HHCB003
Corner Block

1885-1915
Victorian

$4^1/_2$ x $4^1/_2$
or $5^1/_2$ x $5^1/_2$

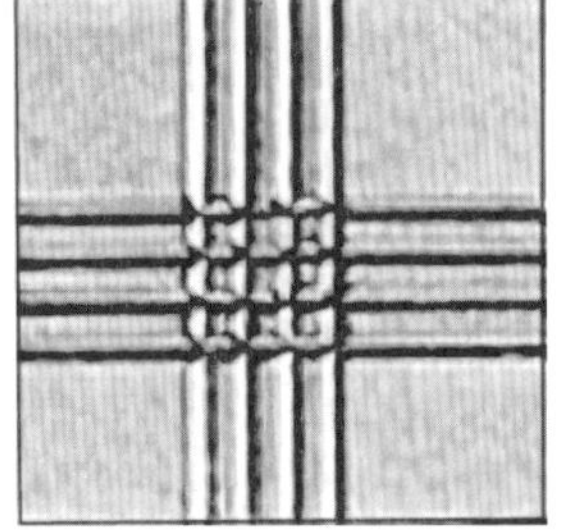

HHCB004
Corner Block

1885-1915
Victorian

$4^1/_2$ x $4^1/_2$
or $5^1/_2$ x $5^1/_2$

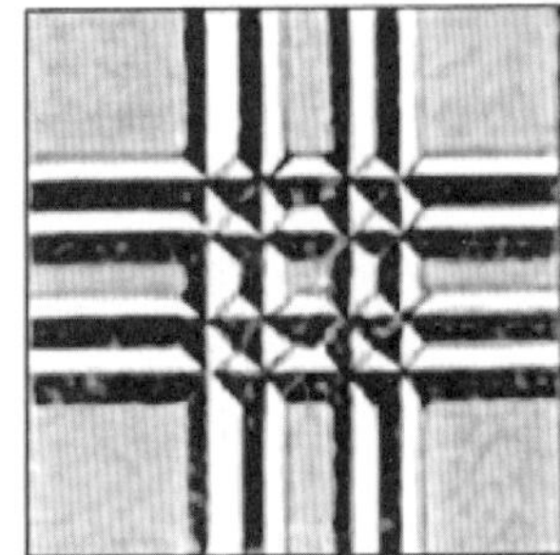

HHCB005
Corner Block

1885-1915
Victorian

$4^1/_2$ x $4^1/_2$
or $5^1/_2$ x $5^1/_2$

HHCB006
Corner Block

1885-1915
Victorian

$4^1/_2$ x $4^1/_2$
or $5^1/_2$ x $5^1/_2$

HHCB007
Corner Block

1885-1915
Victorian

$4^1/_2$ x $4^1/_2$
or $5^1/_2$ x $5^1/_2$

HHCB008
Corner Block

1913
Victorian

$4^1/_2$ x $4^1/_2$
or $5^1/_2$ x $5^1/_2$

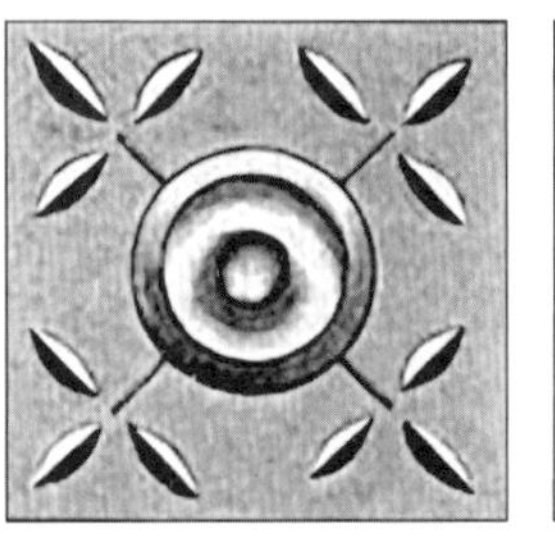
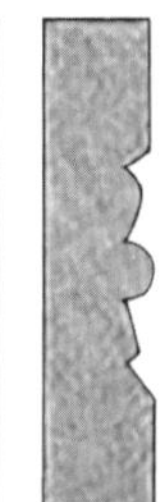

HHCB009
Corner Block

1885-1915
Victorian

$4^1/_2$ x $4^1/_2$
or $5^1/_2$ x $5^1/_2$

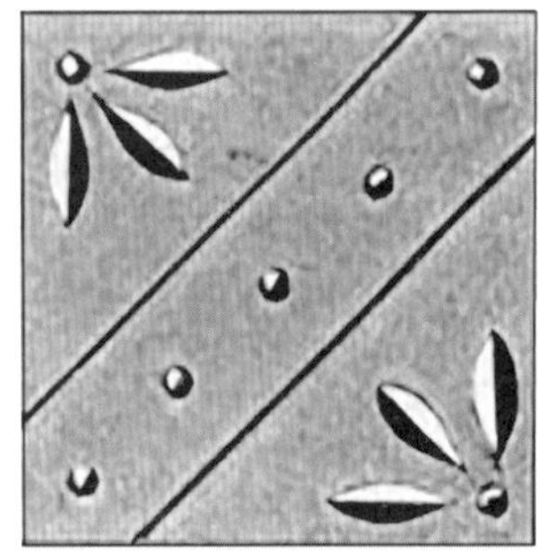

HHCB010
Corner Block

1885-1915
Victorian

$4^1/_2$ x $4^1/_2$
or $5^1/_2$ x $5^1/_2$

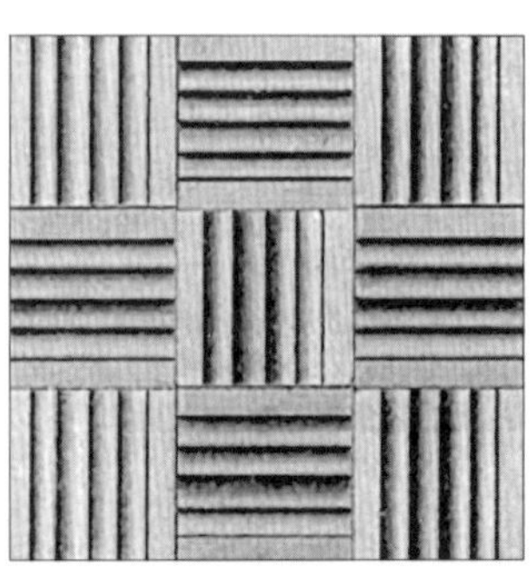

HHCB011
Corner Block
1885-1915
Victorian

4 1/2 x 4 1/2
or 5 1/2 x 5 1/2

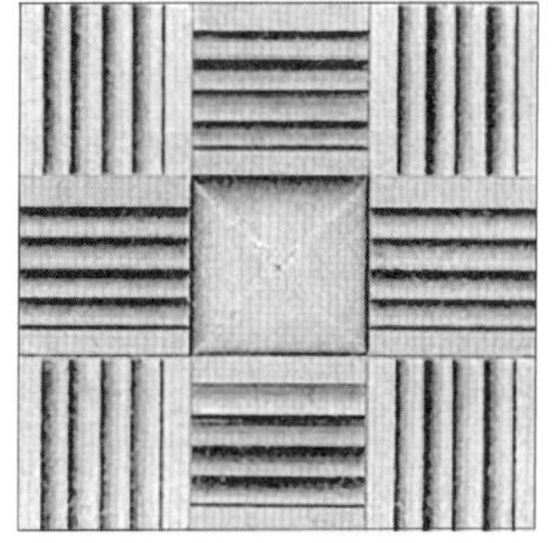

HHCB012
Corner Block
1885-1915
Victorian

4 1/2 x 4 1/2
or 5 1/2 x 5 1/2

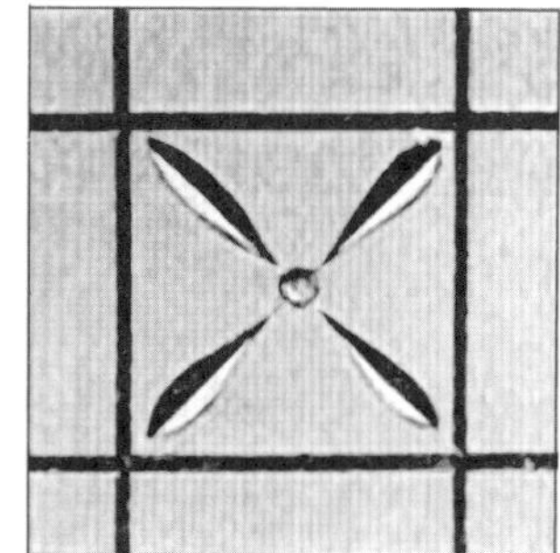

HHCB013
Corner Block
1885-1915
Victorian

4 1/2 x 4 1/2
or 5 1/2 x 5 1/2

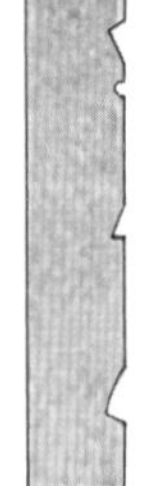

HHCB014
Corner Block
1885-1915
Victorian

4 1/2 x 4 1/2
or 5 1/2 x 5 1/2

HHCB015
Corner Block
1885-1915
Victorian

4 1/2 x 4 1/2
or 5 1/2 x 5 1/2

HHCB016
Corner Block
1885-1915
Victorian

4 1/2 x 4 1/2
or 5 1/2 x 5 1/2

HHCB017
Corner Block
1885-1915
Victorian

4 1/2 x 4 1/2
or 5 1/2 x 5 1/2

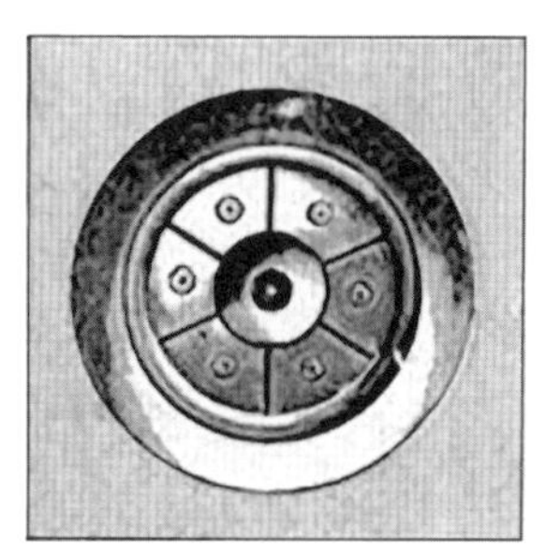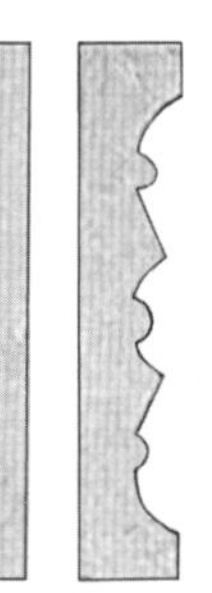

HHCB018
Corner Block
1885-1915
Victorian

4 1/2 x 4 1/2
or 5 1/2 x 5 1/2

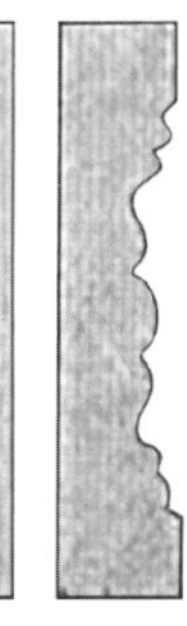

HHCB019
Corner Block

1885-1915
Victorian

$4^1/_2$ x $4^1/_2$

or $5^1/_2$ x $5^1/_2$

HHCB020
Corner Block

1885-1915
Victorian

$4^1/_2$ x $4^1/_2$

or $5^1/_2$ x $5^1/_2$

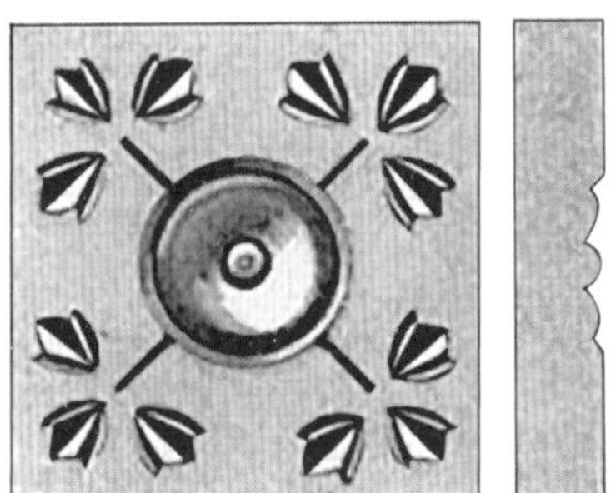

HHCB021
Corner Block

1885-1915
Victorian

$4^1/_2$ x $4^1/_2$

or $5^1/_2$ x $5^1/_2$

HHCB022
Corner Block

1885-1915
Victorian

$4^1/_2$ x $4^1/_2$

or $5^1/_2$ x $5^1/_2$

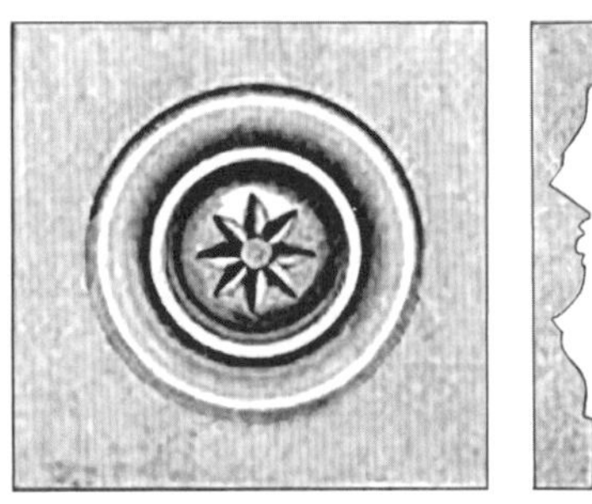

HHCB023
Corner Block

1885-1915
Victorian

$4^1/_2$ x $4^1/_2$

or $5^1/_2$ x $5^1/_2$

HHCB024
Corner Block

1885-1915
Victorian

$4^1/_2$ x $4^1/_2$

or $5^1/_2$ x $5^1/_2$

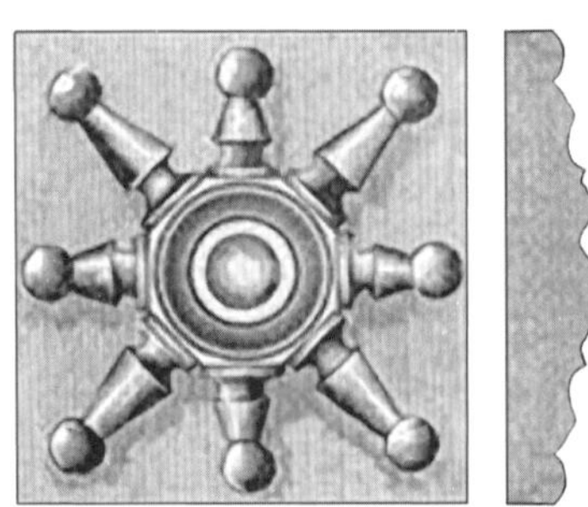

HHCB25
Corner Block

1885-1915
Victorian

$5^1/_2$ x $5^1/_2$

HHCB026
Corner Block

1885-1915
Victorian

$5^1/_2$ x $5^1/_2$

HHCB027
Corner Block
1885-1915
Victorian

4¹/₂ x 4¹/₂

or 5¹/₂ x 5¹/₂

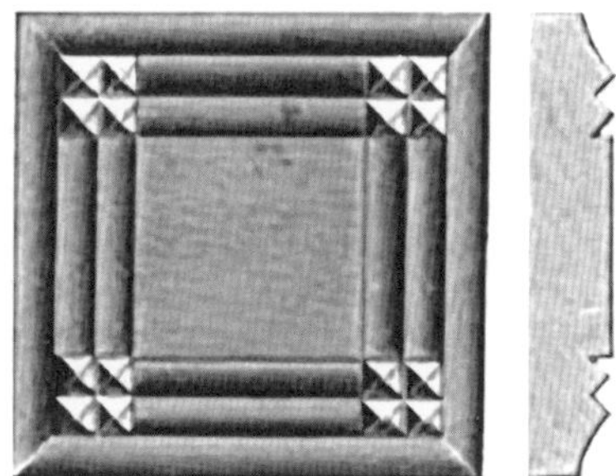

HHCB028
Corner Block
1885-1915
Victorian

4¹/₂ x 4¹/₂

or 5¹/₂ x 5¹/₂

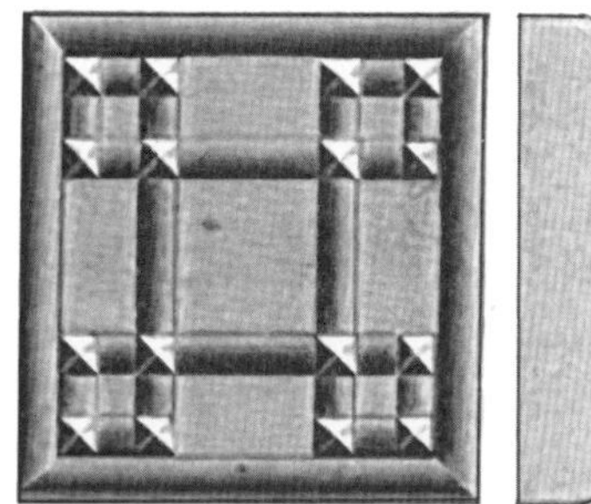

HHCB029
Corner Block
1885-1915
Victorian

4¹/₂ x 4¹/₂

or 5¹/₂ x 5¹/₂

HHCB030
Corner Block
1885-1915
Victorian

4¹/₂ x 4¹/₂

or 5¹/₂ x 5¹/₂

HHCB031
Corner Block
1885-1915
Victorian

4¹/₂ x 4¹/₂

or 5¹/₂ x 5¹/₂

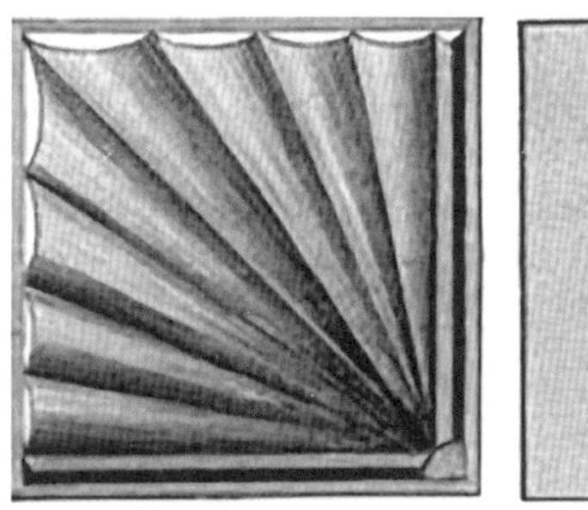

HHCB032
Corner Block
1885-1915
Victorian

4¹/₂ x 4¹/₂

or 5¹/₂ x 5¹/₂

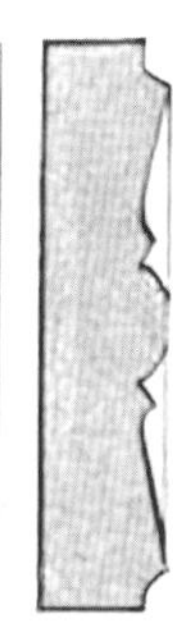

HHCB033
Corner Block
1885-1915
Victorian

4¹/₂ x 4¹/₂

or 5¹/₂ x 5¹/₂

HHCB034
Corner Block
1885-1915
Victorian

4¹/₂ x 4¹/₂

or 5¹/₂ x 5¹/₂

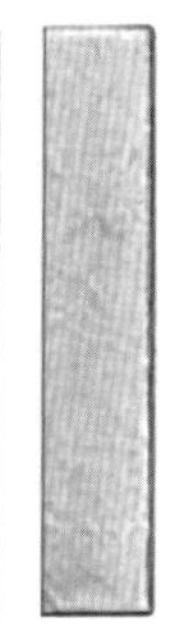

HHCB0335
Corner Block
1885-1915
Victorian

$4^{1}/_{2}$ x $4^{1}/_{2}$

or $5^{1}/_{2}$ x $5^{1}/_{2}$

HHCB036
Corner Block
1885-1915
Victorian

$4^{1}/_{2}$ x $4^{1}/_{2}$

or $5^{1}/_{2}$ x $5^{1}/_{2}$

HHCB037
Corner Block
1885-1915
Victorian

$4^{1}/_{2}$ x $4^{1}/_{2}$

or $5^{1}/_{2}$ x $5^{1}/_{2}$

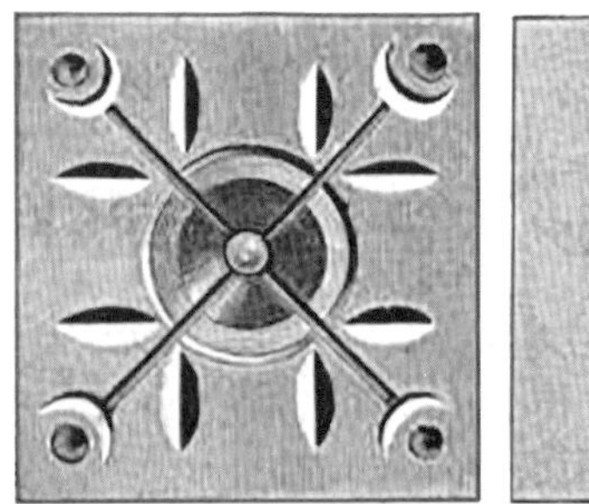

HHCB038
Corner Block
1885-1915
Victorian

$4^{1}/_{2}$ x $4^{1}/_{2}$

or $5^{1}/_{2}$ x $5^{1}/_{2}$

HHCB039
Corner Block
1885-1915
Victorian

$4^{1}/_{2}$ x $4^{1}/_{2}$

or $5^{1}/_{2}$ x $5^{1}/_{2}$

HHCB040
Corner Block
1885-1915
Victorian

$4^{1}/_{2}$ x $4^{1}/_{2}$

or $5^{1}/_{2}$ x $5^{1}/_{2}$

HHCB041
Corner Block
1885-1915
Victorian

$4^{1}/_{2}$ x $4^{1}/_{2}$

or $5^{1}/_{2}$ x $5^{1}/_{2}$

HHCB042
Corner Block
1885-1915
Victorian

$4^{1}/_{2}$ x $4^{1}/_{2}$

or $5^{1}/_{2}$ x $5^{1}/_{2}$

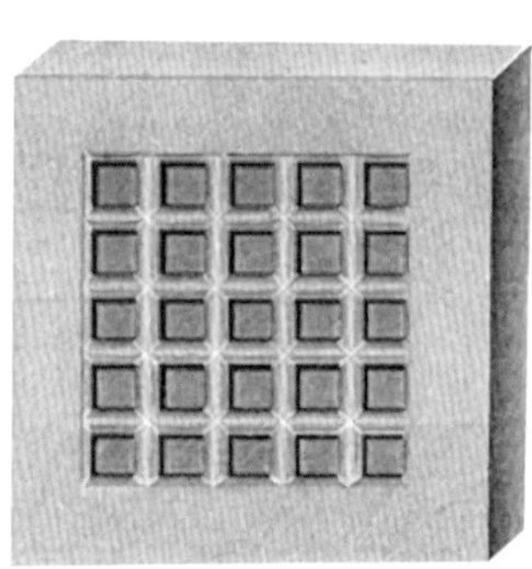

HHCB043
Corner Block

1885-1915
Victorian

$4^1/_2$ x $4^1/_2$
$5^1/_2$ x $5^1/_2$

HHCB044
Corner Block

1885-1915
Victorian

$4^1/_2$ x $4^1/_2$
$5^1/_2$ x $5^1/_2$

HHHB001
Head Block

1885-1915
Victorian

$5^1/_2$ x 10

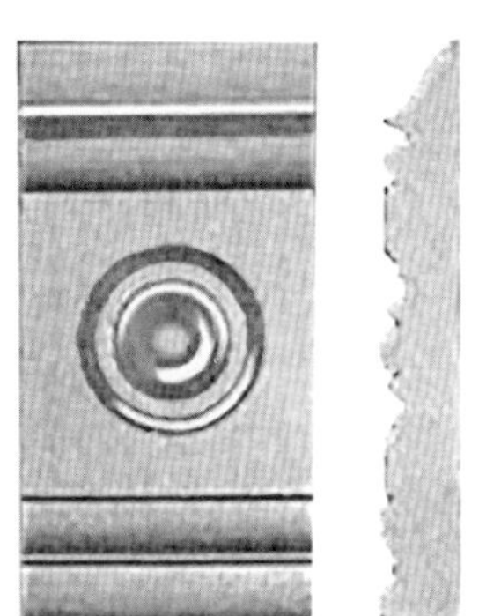

HHHB002
Head Block

1885-1915
Victorian

$5^1/_2$ x 10

HHHB003
Head Block

1885-1915
Victorian

$5^1/_2$ x 10

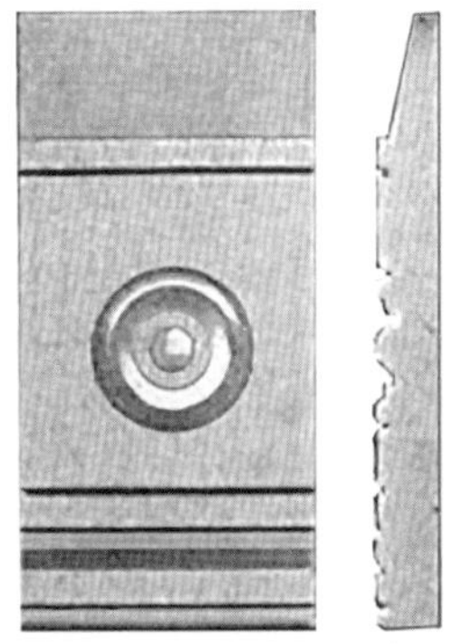

HHHB004
Head Block

1885-1915
Victorian

$5^1/_2$ x 10

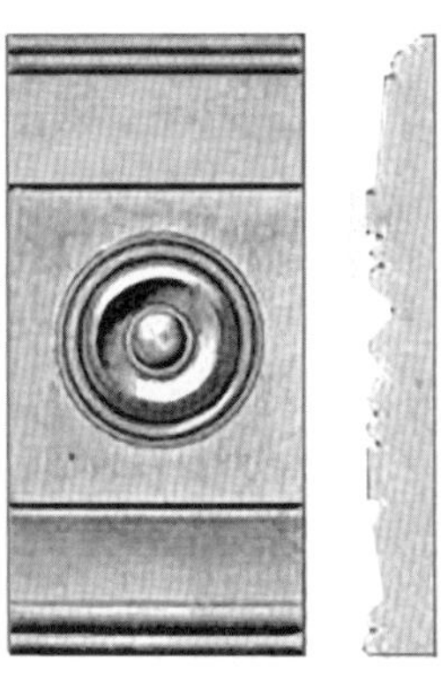

HHHB005
Head Block

1885-1915
Victorian

$5^1/_2$ x 11

HHHB006
Head Block

1885-1915
Victorian

$5^1/_2$ x 11

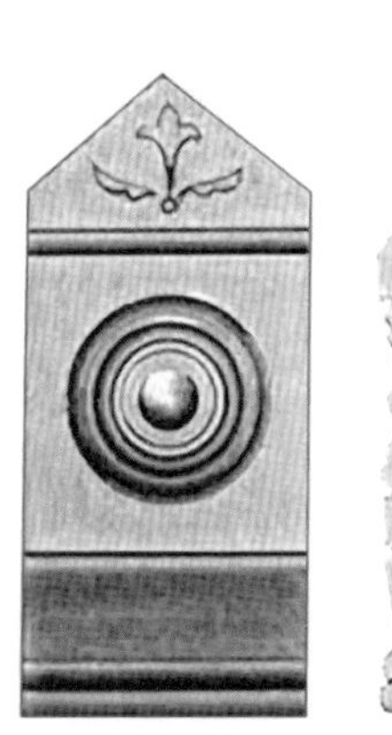

HHHB007
Head Block
1885-1915
Victorian

$5^{1}/_{2}$ x 10

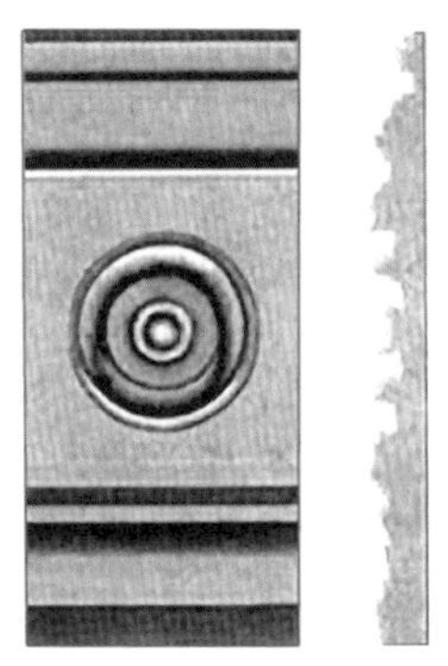

HHHB008
Head Block
1913
Victorian

$5^{1}/_{2}$ x 10

HHHB009
Head Block
1885-1915
Victorian

$5^{1}/_{2}$ x 12

HHHB010
Head Block
1885-1915
Victorian

$5^{1}/_{2}$ x 12

HHHB011
Head Block
1885-1915
Victorian

$5^{1}/_{2}$ x 11

HHHB012
Head Block
1885-1915
Victorian

$5^{1}/_{2}$ x 11

HHHB013
Head Block
1885-1915
Victorian

$5^{1}/_{2}$ x 11

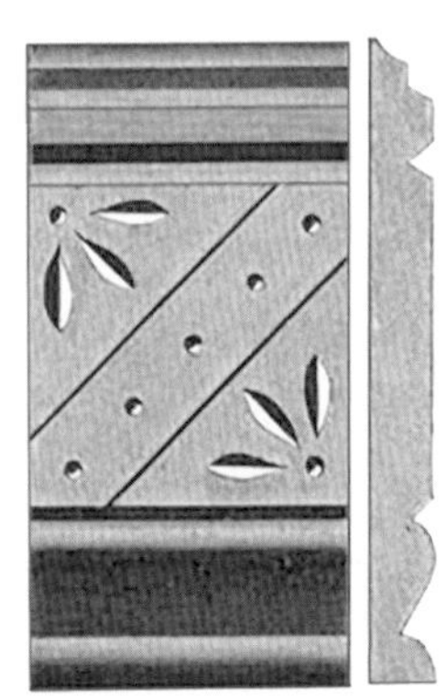

HHHB014
Head Block
1885-1915
Victorian

$5^{1}/_{2}$ x 10

HHHB015
Head Block
1885-1915
Victorian

$5^1/_2$ x 10

HHHB016
Head Block
1885-1915
Victorian

$5^1/_2$ x 12

HHHB017
Head Block
1885-1915
Victorian

$5^1/_2$ x 12

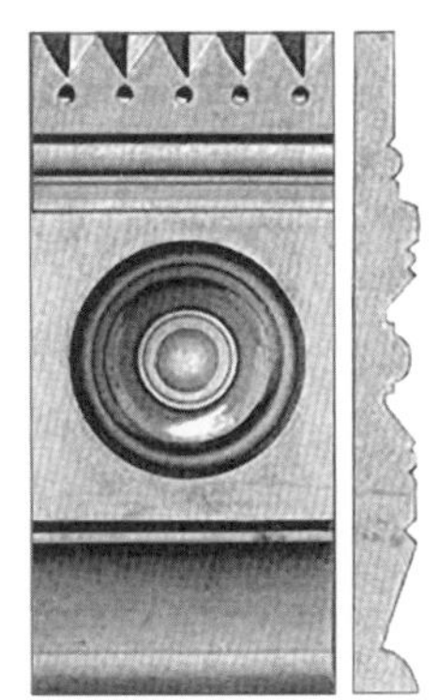

HHHB018
Head Block
1885-1915
Victorian

$5^1/_2$ x $9^1/_2$

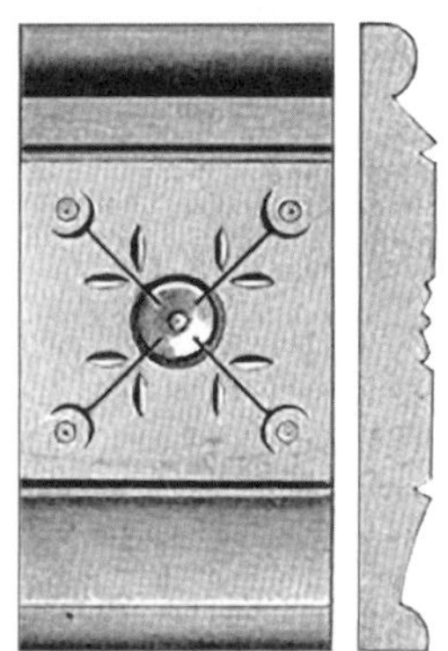

HHHB019
Head Block
1885-1915
Victorian

$5^1/_2$ x 10

HHHB020
Head Block
1885-1915
Victorian

$5^1/_2$ x 10

HHHB021
Head Block
1885-1915
Victorian

$5^1/_2$ x 10

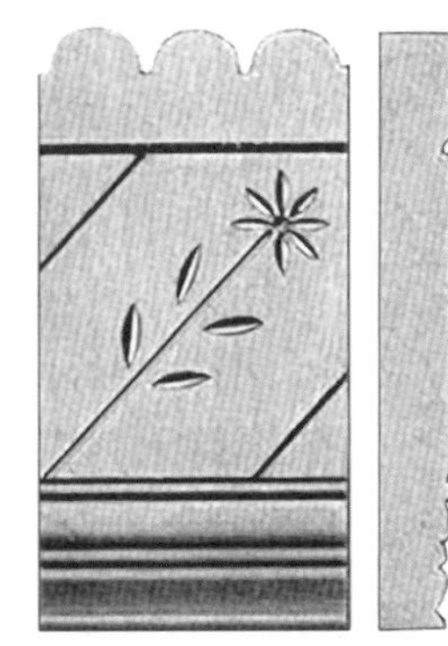

HHHB022
Head Block
1885-1915
Victorian

$5^1/_2$ x 11

HHHB023
Head Block
1885-1915
Victorian

3 x 3

HHHB024
Head Block
1885-1915
Victorian

3 x 3

HHHB025
Head Block
1885-1915
Victorian

$5^1/_2$ x $5^1/_2$

HHHB026
Head Block
1885-1915
Victorian

$5^1/_2$ x 11

HHHB027
Head Block
1885-1915
Victorian

$5^1/_2$ x $10^1/_2$

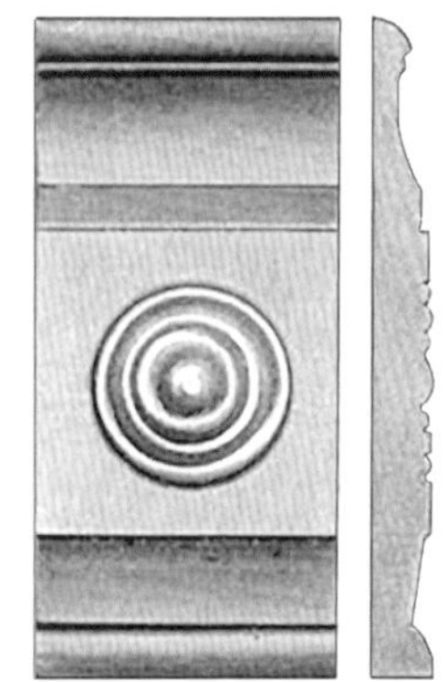

HHHB028
Head Block
1885-1915
Victorian

$5^1/_2$ x 10

HHHB029
Head Block
1885-1915
Victorian

$5^1/_2$ x 11

HHHB030
Head Block
1885-1915
Victorian

$5^1/_2$ x $11^1/_2$

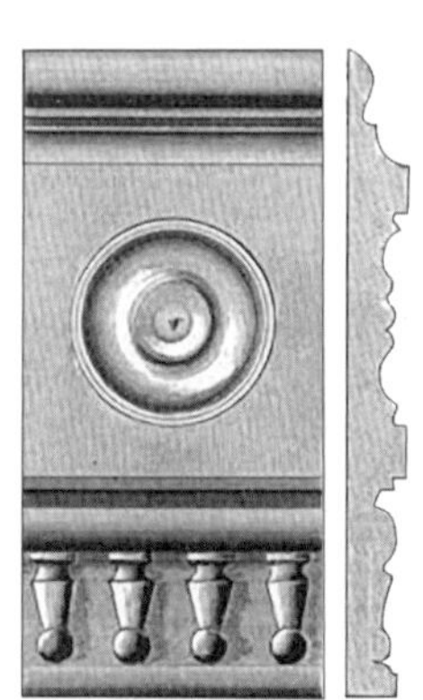

HHHB031
Head Block

1885-1915
Victorian

$5^1/_2$ x 11

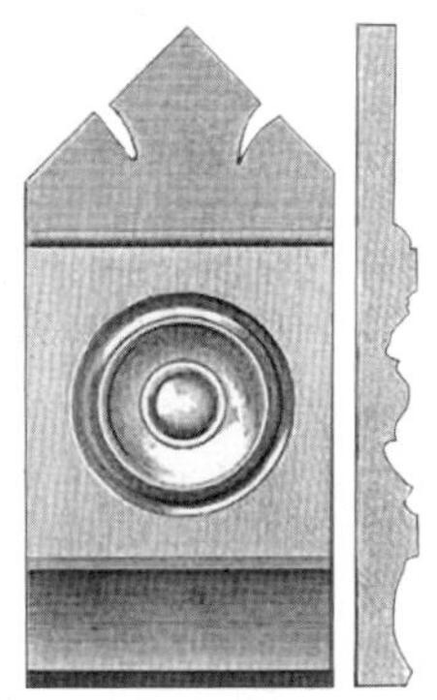

HHHB032
Head Block

1885-1915
Victorian

$5^1/_2$ x $10^1/_2$

HHHB033
Head Block

1885-1915
Victorian

5 — $5^1/_2$ x 10

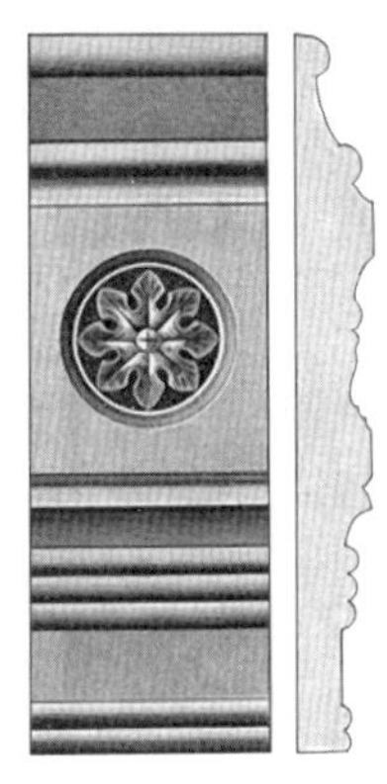

HHHB034
Head Block

1885-1915
Victorian

$5^1/_2$ x $5^1/_2$

HHHB035
Head Block

1885-1915
Victorian

$5^1/_2$ x 9

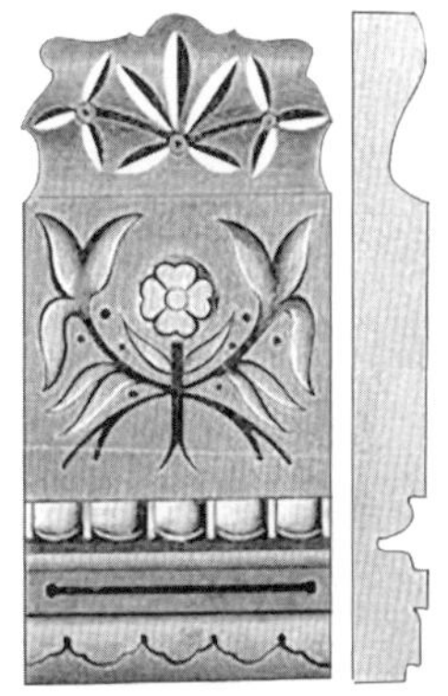

HHHB036
Head Block

1885-1915
Victorian

$5^1/_2$ x $12^1/_2$

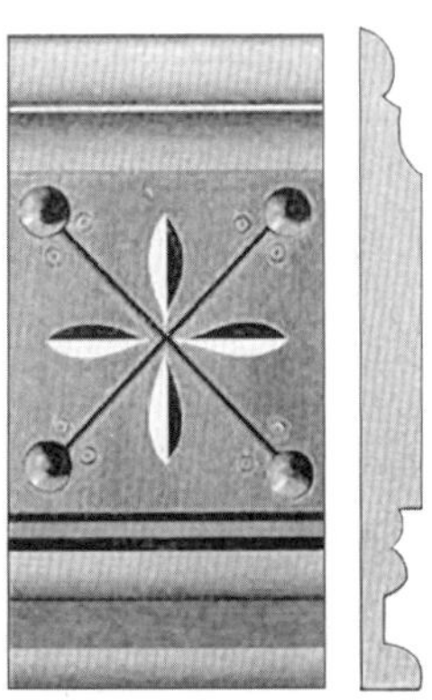

HHHB037
Head Block

1885-1915
Victorian

$5^1/_2$ x $11^1/_2$

HHHB038
Head Block

1885-1915
Victorian

$5^1/_2$ x $5^1/_2$

HHHB039
Head Block

1885-1915
Victorian

3 x 3

HHHB040
Head Block

1885-1915
Victorian

3 x 3

HHHB041
Head Block

1885-1915
Victorian

5^1/$_2$ x 12

HHHB042
Head Block

1885-1915
Victorian

5^1/$_2$ x 12

HHHB043
Head Block

1885-1915
Victorian

5^1/$_2$ x 12^1/$_2$

PANEL MOLDINGS AND SUNK PANEL MOLDINGS

panel mold

HULL
HISTORICAL
MOLDING

Panel Moldings

There may be no more confusing portion of historic molding catalogs than the panel molds. The variety of sizes and types is simply dazzling to the modern eye. In one catalog I saw over 17 varieties of the exact same molding. Each of the 17 was a different size depending on whether you planned to use it as a rabbeted panel mold, a base cap, a sunk mold or something else. But these weren't the only uses. If you've studied Victorian or other historic homes, you know that sometimes local traditions prevailed and a panel mold was used in some unlikely area, like the cornice or outside fascia.

Most of the panel molds are Victorian and come from the first 1890's Universal molding catalog. In that catalog they were shown together as panel and base cap moldings. We have separated them and the base caps can now be found in the Base Moldings section.

You should also notice that some panel molds are square on the back and others have a notch (rabbet) cut out. These variations permitted the mold to be used in several ways. They could sit flat on the panel with the square edge to the style and rail. They could also lip over the stile and rail before sitting on the panel. Finally, they could be used as sunk molds, sitting below (or sunk below) the face of the stile and rail. You will find all three here.

HHPA001
Panel Molding

1870-1925
Classical Victorian

$1^3/_8$ x 3

This was the most popular style of panel mold and it lasted in this form and size into the 1920s. Simple and classical, the bull nosed edge sits above the face of the door and paneled wall for a great look. You can even use it as a base cap if you like.

HHPA002
Panel Molding

1870-1925
Classical Victorian

$^3/_4$ x 2

This panel mold has a little stronger definition than HHPM001. It too can function as a base cap. I have called it Classical Victorian because it looks Victorian but actually outlives Victorian architecture, lasting until the 1925 catalog.

HHPA003
Panel and Base Molding

1890-1900

$^7/_8$ x $2^1/_4$

I love the OG detail on the front edge of this molding. It gives it a classic flair that would look great painted or stained. The size is typical and could jazz up a panel or base in a new or historic home.

HHPA004
Panel Molding

1890-1915
Victorian

$1^3/_8$ x $4^1/_2$

This is a monster panel mold. Not a common piece but still found in the catalogs. It was most likely used on a section of the house where the scale was grander. If you find a use for this one, let me know —I can't wait to hear about it!

HHPA005
Rabbeted Panel Molding

1880-1900
Victorian

$^7/_8$ x $2^1/_4$

There were almost as many rabbeted (or lipped) panel molds as flat. This was a popular detail on Victorian doors. The term "molded" meant it had a panel mold.

HHPA006
Rabbeted Panel Molding

1880-1900
Victorian

$^7/_8$ x $1^5/_8$

An angular and interesting panel mold that would look great on any new door today. The fillet detail is striking and balances the short width of the molding.

HHPA007
Rabbeted Panel Molding

1880-1900
Victorian

$^7/_8$ x $1^3/_4$

Another shortened version, but handsome. The lip on the back is relatively low so this will project out quite boldly into a room.

HHPA008
Rabbeted Panel Molding

1880-1900
Victorian

$^7/_8$ x 3

A longer and more typical panel mold. Remember that Victorian panels were quite wide, often three inches or more. This panel mold transitioned the stile or rail to the panel quite nicely.

HHPA009
Rabbeted Panel Molding

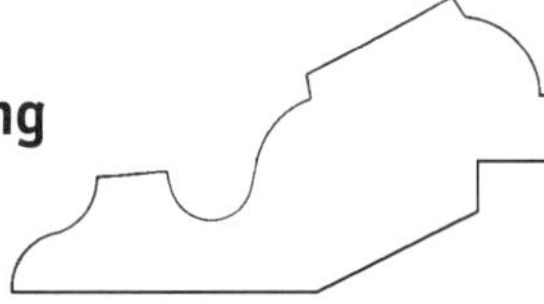

1880-1900
Victorian

$^7/_8$ x $2^1/_2$

My favorite panel mold. We used this on a major courthouse renovation here in Texas and I can vouch for its great looks.

Sunk Panel Moldings

I love these moldings. They are most definitely classic and reminds me of my days in Boston studying historic preservation at the North Bennet St. School. Some of the doors and paneled walls were finished out with a simple sunk mold like this. This look just seems to add a timeless quality to millwork. These sunk molds are mostly Victorian and hence are quite ornate.

HHSP001
Sunk Panel

1890's
Victorian

$1^1/_8$ x $2^3/_4$

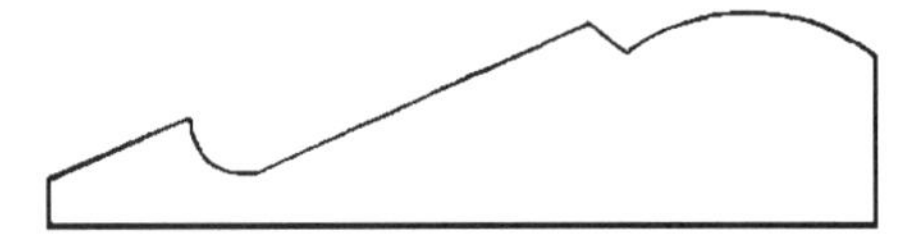

The long projection is typical of many Victorian sunk molds. It has clean lines and is subtle compared to others in this section (if it's even possible at all for a Victorian mold to be subtle!).

HHSP003
Sunk Panel

1890's
Victorian

$1^1/_8$ x $2^5/_8$

Of the long sunk panel molds I like this one the best. It has a subtle bead detail on the inside edge and a raised bead on outside edge that looks great painted.

HHSP004
Sunk Panel

1890's
Victorian

$^7/_8$ x $2^3/_4$

A very bold panel mold that is a bit bigger in size and meant for larger panel details where it can still sit below the face.

HHSP005
Sunk Panel

1890's
Victorian

$1^1/_2$ x $3^3/_4$

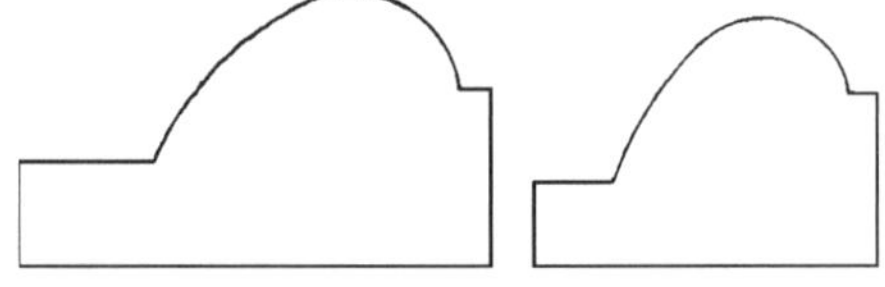

A simple and clean panel mold with a gothic feel. It would look great trimming out a paneled wainscot.

HHSP006
Sunk Panel

1890's
Victorian

$^7/_8$ x $2^3/_4$
$1^1/_8$ x $2^5/_8$

We offer this molding in two sizes. Similar to HHSP005, but with a cove detail in the tail or top. This is a neat sunk mold that can really add character to your historic or new project.

HHSP007
Sunk Panel

1890's
Victorian

$^7/_8$ x $2^3/_4$
$1^1/_8$ x $2^5/_8$

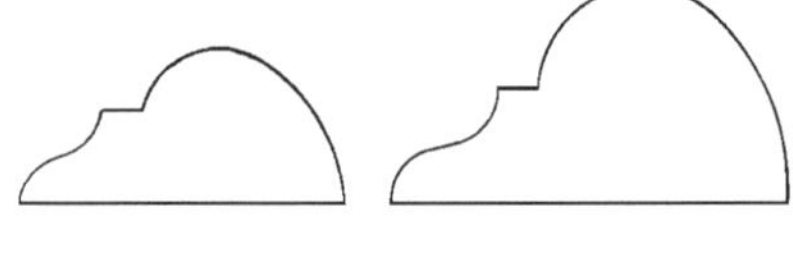

Another great sunk panel mold. The OG detail is just right for a pre-1900 home. I like how the left version really sits low. It is short but well stated.

Miscellaneous

PLATE RAILS, BRICK MOLDS, SCREEN MOLDS, ETC.

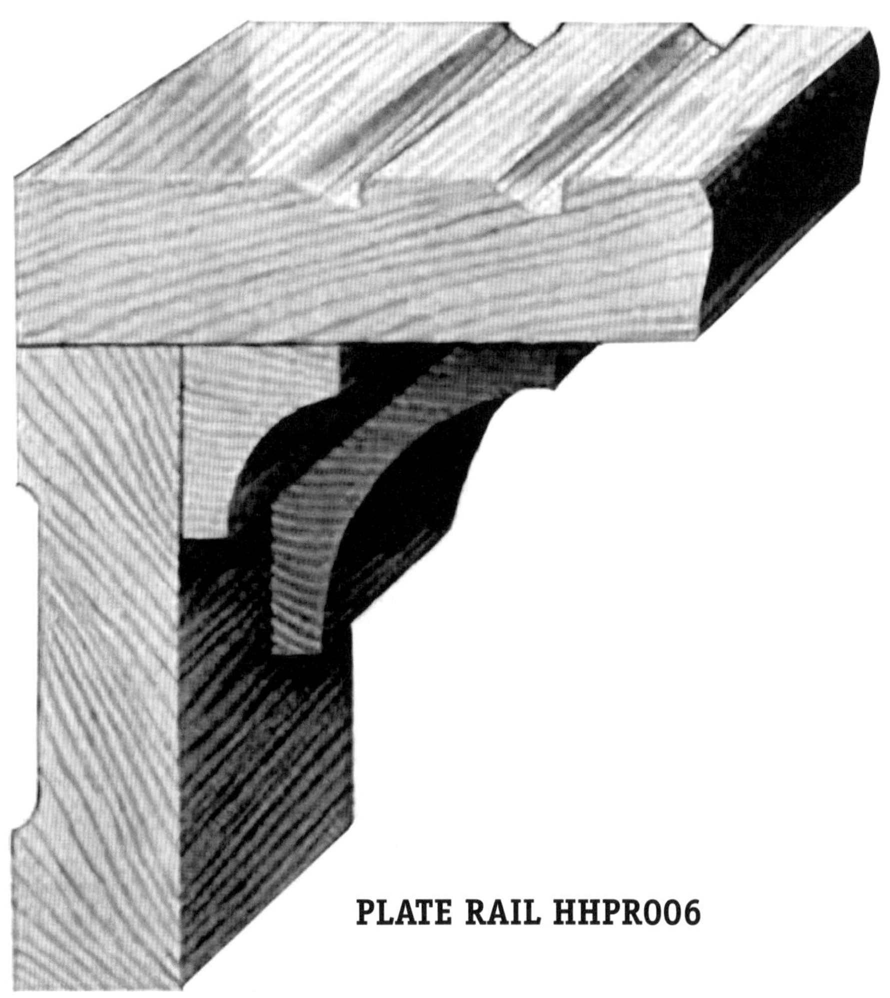

PLATE RAIL HHPR006

HULL
HISTORICAL
MOLDING

Plate Rails

There may be no other turn of the century molding that captures the charm and appeal of historic homes like the plate rail. Usually used in conjunction with some form of wainscoting in the dining room, the plate rail is a decorative trim piece that helps define this period.

Plate rails were not seen in the catalogs until 1900. These moldings were not part of the standard catalog. Plate rails were commonly stained. Our poplar stains beautifully, but we can also run these for you in oak, cherry, or walnut. These plate rails will be shipped in pieces but can be pre-assembled for your convenience at an extra charge.

HHPR001
Plate Rail
1900-1925
Arts & Crafts

projection 4 3/4 in., height 3 3/4 in.

A simple, classic look. The extra bracket adds heft and charm and would look great in a historic or new home. If moved up to the door header heights, it will give a great cornice.

HHPR003
Plate Rail
1900-1915
Arts & Crafts

projection 3 3/4 in., height 2 3/4 in.

This short and thick plate rail is a three-slotted rail with simpler details.

HHPR004
Plate Rail
1900-1915
Arts & Crafts

projection 5 1/4 in., height 5 3/4 in.

Another simple plate rail that uses the decorative bracket. The bracket stands alone and should be spaced 12-18 inches on center.

HHPR005
Plate Rail
1910-1925
Arts & Crafts

projection 4 1/2 in., height 5 1/4 in.

This relatively unornamented plate rail would look great in a Craftsman or other simple home from the turn of the century. It could even work in a modern home where clean, sleek lines are desirable. I like the square edges and cove mold.

HHPR006
Plate Rail
1910-1925
Arts & Crafts/Period Revival

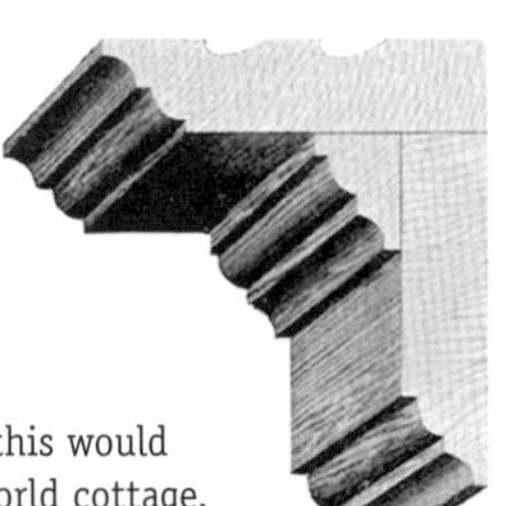

projection 4 in., height 4 in.

The built-up plate rail with two grooves, this would go well in a Period Revival home or Old World cottage.

HHPR007
Plate Rail
1910-1925
Arts & Crafts

projection 3 3/4 in., height 4 in.

This is my favorite plate rail. It has clean lines and a great compact bracket. The simple design makes it work in a Craftsman or Bungalow home, or even in a new traditionally-styled home.

Brick Moldings

As a custom provider of milled windows and doors for historic buildings, one of our side benefits is the discovery of brick molds we have not previously seen. Though many of the brick molds before 1900 are simple quarter rounds in various sizes, those we find after the turn of the century are much more interesting.

The three below are good examples of well-designed brick moldings from 100 years ago. All our brick molds are made from short-leaf yllow pine, a good exterior grade wood for trim. We never use finger-jointed stock for this molding. If you are in a climate with more extreme weather, we can provide it in mahogany or other more durable wood. Brick molds come in 10-foot lengths. Please let us know if you need other lengths.

HHBR001
Brick Molding
1900-1925
Turn Of The Century

$1^3/_4$ x 2

This is my favorite brick mold. We have seen it on buildings dating from the 1890's, but it doesn't show up in the catalogs until 1900. A great-looking mold, it paints well and really highlights the windows and doors.

HHBR002
Brick Molding
1920-1945
Classical Revival

$1^3/_4$ x 2

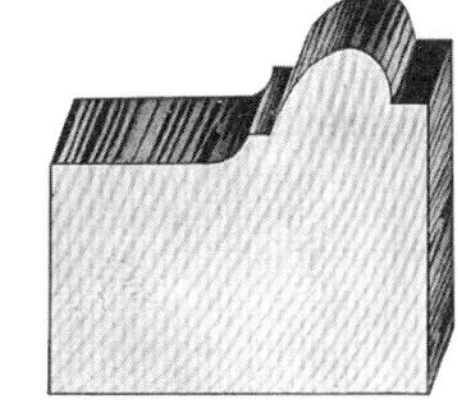

This brick mold has looked wonderful on the buildings where we have used it. Nice shadow lines and classical detailing make it appealing.

HHBR003
Brick Molding
1920-1945
Classical Revival

$1^1/_8$ x 2

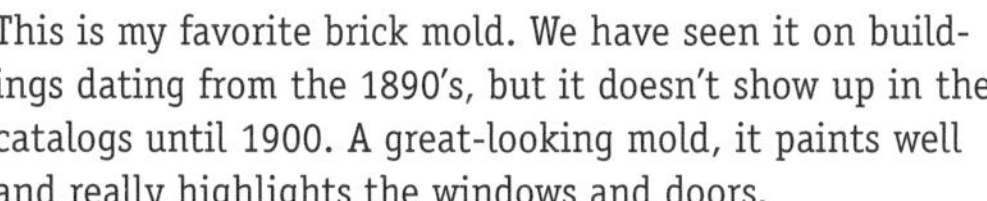

Probably the most popular brick mold from the early 1900's up to the 1950's. The inside bead detail is simple, clean and charming. Go into any historic neighborhood from the 20's and 30's, and you can see this was the brick mold of choice. The back cut detail allows it to tie into the building cleanly.

Screen Moldings

There's nothing like a wood screen door or wood screen window for charm and function uniquely combined. These screen molds are great ways to cover the screen and add distinction to your home.

Some of these screen molds date back to Victorian times, when ornamentation and charm were better understood. Point it out to your friends or clients when they come over—these are great fun.

HHSCM001
Screen Molding

1870-1925
Classical

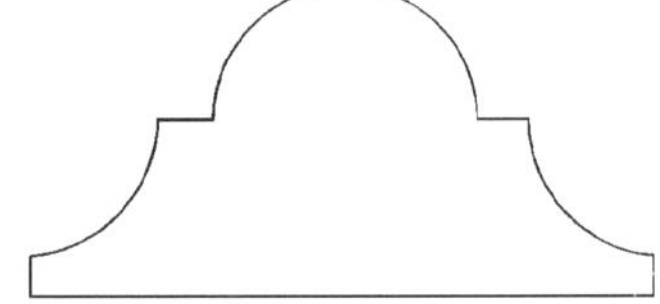

$^3/_8$ x $^3/_4$

This screen mold will really accent your screen doors and windows and beautify your front door. It appeared in catalogs from the 1870's right into the 40's. It's not as common as other types after the 1900's, but is still a great detail.

HHSCM002
Screen Molding

1870-1945
Classical

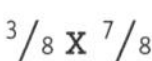

$^3/_8$ x $^7/_8$

We've used this screen mold on a number of Victorian screens and it makes a handsome addition. I remember when I first saw it on an 1899 Victorian; I was struck by how unique and distinctive it looked.

HHSCM003
Screen Molding

1920-1925
Classical Revival

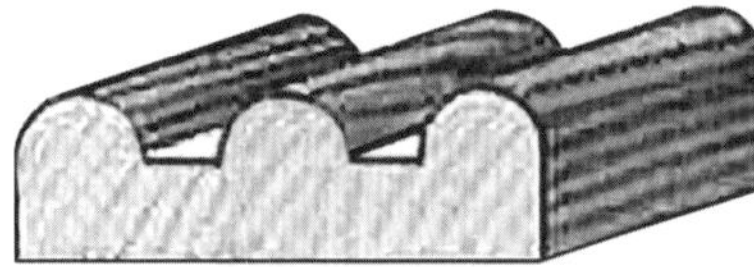

$^1/_4$ x $^3/_4$

This is a wonderful beaded screen mold rarely seen today. It made its debut in the catalogs around 1920 and was popular through the 40's. Interestingly, it has some Victorian detailing, but this speaks to its versatility and ability to look appropriate on any home.

HHSS001
Screen Stock and Slide

1920-1945

a $^1/_2$ x $^3/_4$

b $^3/_4$ x $1^3/_4$

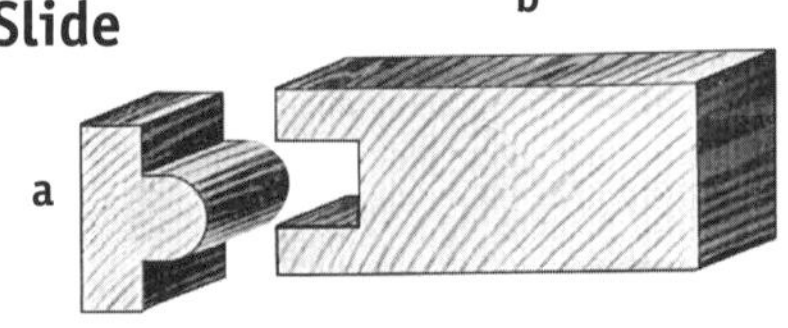

I have put this screen stock in this section because it seems to fit best here. Historic screens were often slotted to slide up out of the way in a double-hung configuration. In this case a screen would be made from only the bottom sash, allowing the screen to slide up out of the way when it needed to be removed.

Wainscot Caps

Historic builders and architects really knew how to make wainscoting. Historic wainscot was lively and charming, not only because wood was cheap and plentiful, but also because the craftsmen of that time were highly skilled. The popularity of the wainscot is reflected by the number of wainscot caps available in the standard catalogs of this time.

Most of these wainscot caps are fashioned to fit over a $^3/_4$-inch wainscot. If yours is deeper, than let us know so that we can provide for your custom depth. These wainscot caps vary from very Classical to very Victorian, but in either case are wonderfully additions to any new or historic home.

HHWC001
Wainscot Cap

1885-1915
Classical

a $1^1/_8$ x $2^1/_2$

b $^3/_4$ x 1

This wainscot cap looks like a window stool with a cove detail underneath. This cap can also work as a picture rail or other decorative edge in certain rooms.

HHWC002
Wainscot Cap

1885-1920
Victorian

$1^1/_8$ x 3

This cap that uses a popular panel mold from Victorian times as a wainscot cap. This speaks for the versatility of the Victorian panel mold, but also can help unify and pull together common details in a new or historic home.

HHWC003
Wainscot Cap

1890-1913
Classical

$1^3/_8$ x $1^3/_4$

This cap is a one-piece version HHWC001. A nice cap that can also function as a picture rail as suggested above.

HHWC004
Wainscot Cap

1900-1915
Late Victorian

a $^{13}/_{16}$ x $2^1/_8$

This very ornate Victorian cap doesn't appear in the standard catalogs until 1900. It has a typical Victorian bead detail on the face and decorative edge. It could adapt well to the plywood sheets of bead board offered by many companies today.

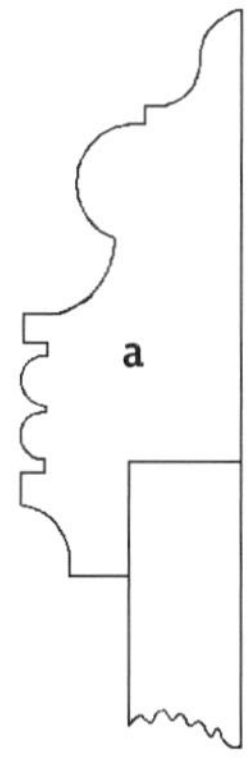

HHWC005
Wainscot Cap

1885-1915

a $^7/_8$ x $2^1/_8$

b $^5/_8$ x $^7/_8$

An angular and interesting cap, this combines some classic Victorian lines. It is great way to dress up the wainscot in any home.

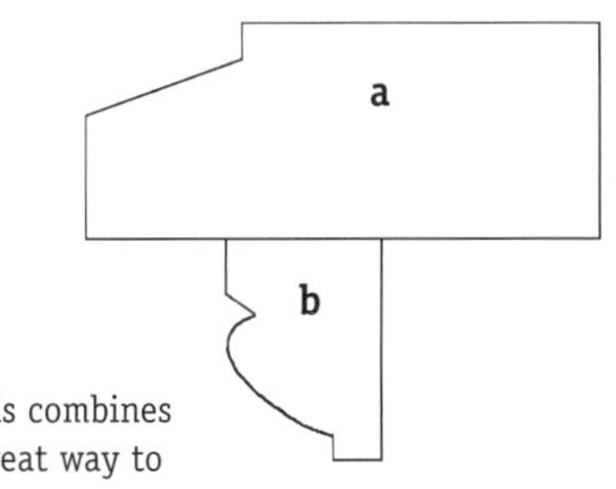

Astragal Moldings

Astragal molds are an archaic yet charming detail that was used historically to highlight or bring attention to moldings and paneling. It was even used as a batten detail on vertical siding. Astragals were a common item in Victorian catalogs and were often used in smaller sizes as screen molds.

Remember the Victorian moldings were often heavily built up and ornate and astragals are part of that heritage. We provide these astragals to allow you to match existing pieces or for new designs you may want to highlight.

HHAS001
Astragal Molding

1870-1920
Classical

$1^3/_8$ x 3

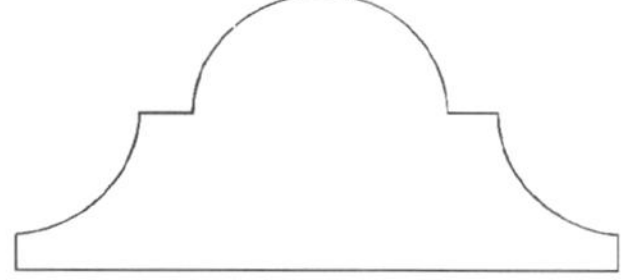

This astragal mold dates back to one the earliest catalogs in 1870 and lasts there until 1920.

HHAS002
Astragal Molding

1885-1900
Victorian

$3/_4$ x $1^1/_2$

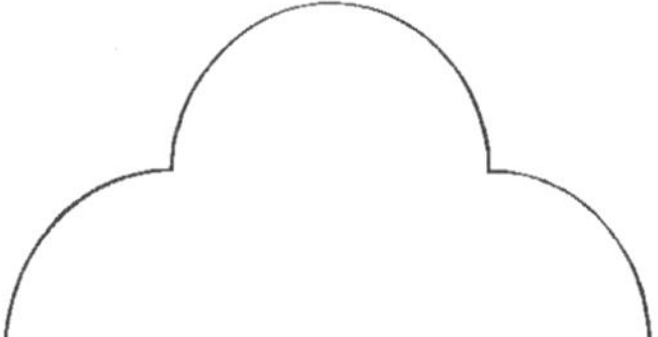

A great looking Victorian astragal that uses the quirked bead to bring almost a gothic highlight to architectural details and trim. This would make an appealing painted highlight detail.

HHAS003
Astragal Molding

1890-1900
Victorian

a $^7/_8$ x 2

b $1^1/_8$ x $2^3/_4$

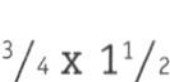

This Victorian astragal was offered in two sizes. They both use the astragal cap with a flanking OG on either side.

HHAS004
Astragal Molding

1890-1915
Classical

$^7/_8$ x $2^1/_4$

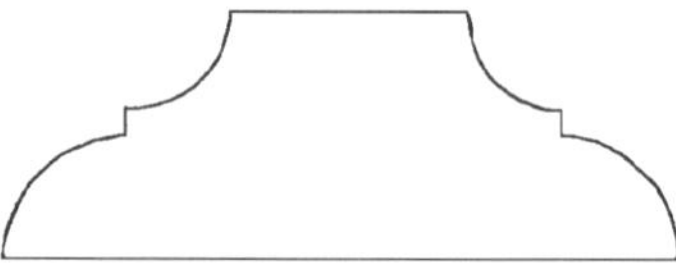

This astragal features an interesting top flanked by a bead and cove detail. This great-looking astragal was popular through two periods.

HHAS005
Astragal Molding

1885-1900
Victorian

$1^1/_8$ x $3^1/_2$

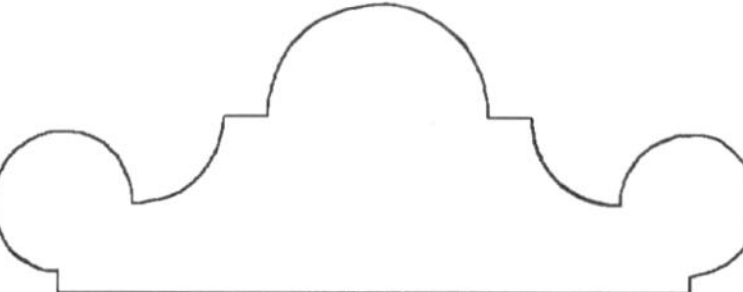

A wild Victorian astragal which is also the largest we offer. This astragal is so big it could even function as a casing in some settings. It is symmetrically oriented and would cap a casing nicely as well.

HHAS006
Astragal Molding

1885-1900
Victorian

$^7/_8$ x $2^1/_4$

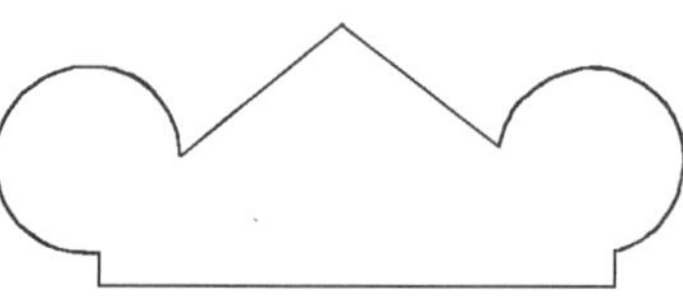

Another bold Victorian astragal, this uses the sharp peak in the middle of the mold for a gothic feel. The quirked bead is appropriate and handsome.

Door Astragals

Today's door astragals are so plain and simple with no ornamentation at all. However, historic door astragals for both swinging and sliding doors were charming and bold. We have been using these astragals on our own projects ever since we found them in the historical catalogs several years ago. We find they create a wonderful division and definition between the doors.

Interestingly, the door astragal isn't a standard catalog item until after 1900. Many doors before then were rabbeted with a bead where they meet. Notice the bead detail on most of our astragals that convey this historic look. All astragals are provided in 8-foot lengths, but other lengths are available. Also be sure to specify the door thickness.

HHDA001
**Astragal Molding
for Folding Doors**

1900-1915
Late Victorian

$1^3/_4$ x $2^1/_2$

The most ornate astragal we offer. Bold and heavily detailed, it fits on an historic or even a new Old World style door. This rich and wonderful astragal is an example of a heavy Victorian design that still looks great today.

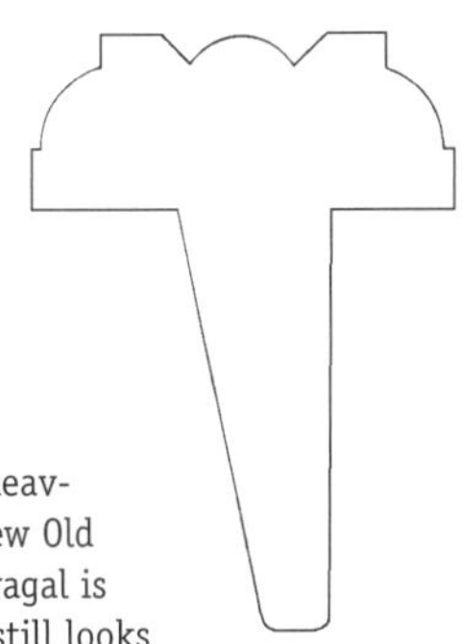

HHDA002
**Astragal Molding
for Folding Doors**

1900-1915
Late Victorian

$1^5/_8$ x $2^7/_8$

This is probably my favorite door astragal—just the right amount of ornamentation topped with an astragal cap.

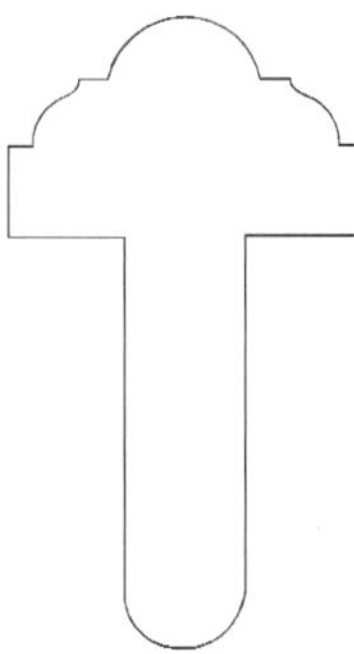

HHDA003
**Astragal Molding
for Folding Doors**

1925-1945
Classical Revival

$1^5/_{16}$ x $2^3/_8$

Door astragals grow simpler with time. These next two astragals are offerings that come from catalogs after 1925. They are simple and probably should be matched up with the type of detail that appears on the door.

HHDA004
**Astragal Molding
for Folding Doors**

1925-1945
classical Revival

$1^1/_{16}$ x $2^3/_8$

Like HHDA003, this astragal is a late type that would work best with a door that has the same bead and cove profile. It is a simple way to tie moldings together and make these entire packages conform.

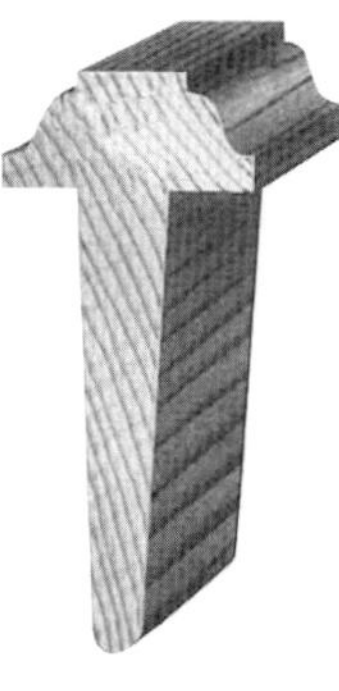

HHDA005
**Astragal Molding
for Sliding Doors**

1900-1930
Classical

a $^{13}/_{16}$ x $2^3/_8$

b $^{13}/_{16}$ x $2^3/_8$

Sliding doors were a popular Victorian detail that separated rooms. Often the astragals joining these doors were decorative, and even the hardware would conform to this odd shape. These two astragals are a great sliding door detail.

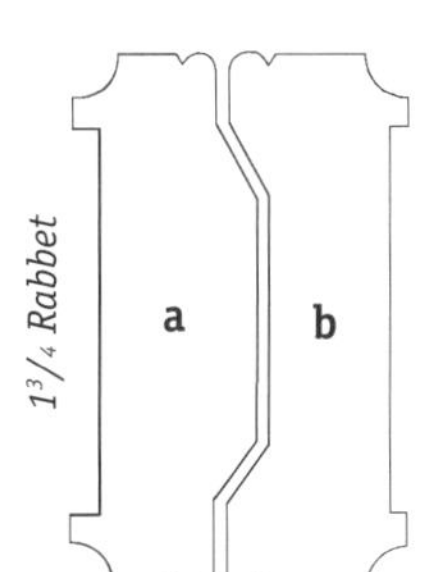

HHDA006
**Astragal Molding
for Sliding Doors**

1900-1915
Late Victorian

a $^{13}/_{16}$ x $2^3/_8$

b $^{13}/_{16}$ x $2^3/_8$

This sliding door astragal uses double-beaded sides and a unique joint to create a nice detail on a door, open or closed.

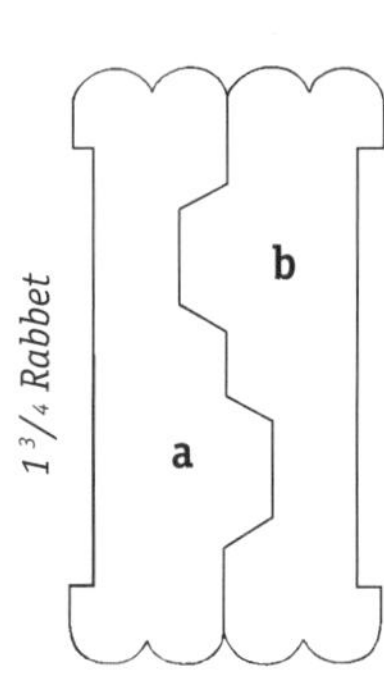

The Best of the Rest

The following moldings are so historically unique that "miscellaneous" is the only suitable description.

For example, we have included the electrical molds that appeared in the standard catalogs in 1900. I doubt these are covered in the International Building Code but they are historical and tell us about ourselves and how far we have come. There is also a blackboard molding that was available in the 20's and lattice which used to be a popular molding item.

In addition, there are also moldings here that must be included to make any historic molding catalog complete. Return and outside beads, drip ledges, nosings and chair rails, just to name a few.

HHBT001
Batten Molding

1870-1915
Victorian

$7/8$ x 2

Batten molds are the dividing strip between two vertical boards on siding or doors. These batten strips are mainly Victorian moldings that were common on wood framed buildings. The molded face of these batten moldings are great definition to any historic home or barn.

HHBT002
Batten Molding

1885-1900

$3/4$ x $5^1/4$

This is a popular batten I have seen many times on vertical siding on homes and barns. A simpler batten, it is still bold and looks great painted.

HHBT003
Batten Molding

1900-1915
Late Victorian

$1/2$ x 2

A late Victorian batten mold that is thinner than earlier versions and could be used on plywood walls to give a wood wall a paneled-out feel.

HHBL001
Blackboard Molding

1920-1935
Classical Revival

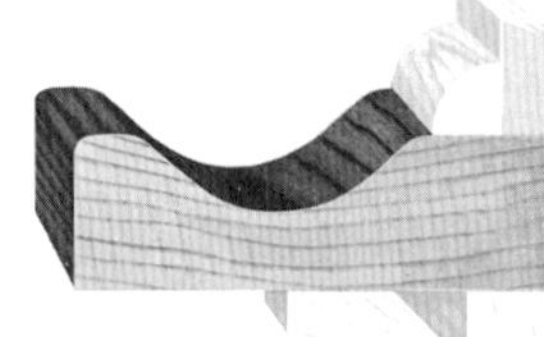

$1^1/13$ x $3^1/2$

Here it is: Resurrected and good as new, this charmer would go great in a kid's room, play room or historic school house restoration. Sold as a single piece, the quarter round and cove mold are also available, but if you would like the whole package, tell us and we can quote it either way.

HHBC001
Beaded Ceiling

1885-1920

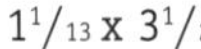

$3/4$ x $3^1/4$

Our beaded ceiling is a little different from some you may have seen, but its unique design can really give a home that special look. This mold was seen in catalogs from the 1890's to 1920, and thus transitions several periods.

HHBC002
Beaded Ceiling

1885-1900

$3/4$ x $6^1/2$

This ceiling mold was originally run as two pieces, but we have combined it for faster installation. A great ceiling when painted, it gives a strong period look to any room.

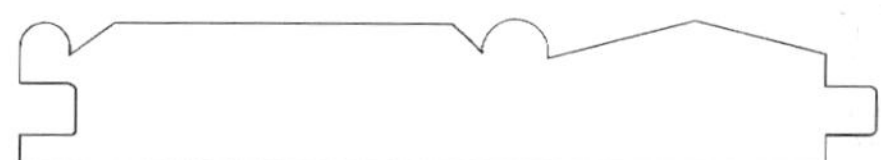

HHBC003
Beaded Ceiling

1885-1900
Victorian

$^7/_8$ x 2

Another wide ceiling board, it makes a great porch ceiling for new or historic homes.

HHCH001
Chair Rail

1920-1945
Period Revival

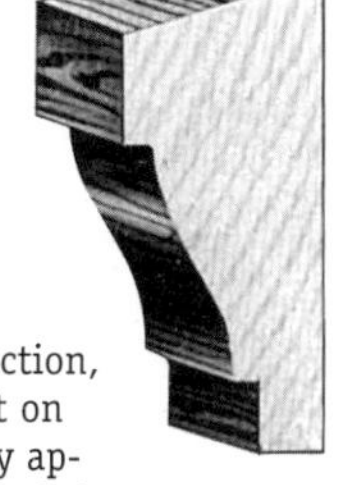

$^7/_8$ x 3$^1/_2$

The chair rail, as you may remember from the Introduction, was inspired from classical architecture and is laid out on the wall proportionately with the pedestal cap. It only appears in the catalogs after the 1920s, taking the place of the wainscot cap as wainscot fell from favor.

HHCH002
Chair Rail

1920-1945
Period Revival

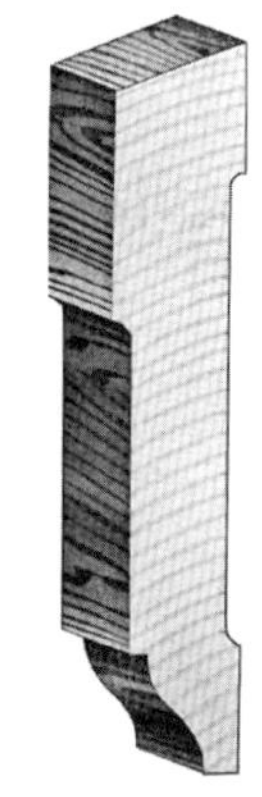

$^7/_8$ x 4$^1/_4$

This is a great chair rail that we have used on many historic restorations. It has as slanted top and breaks up the wall in a proportionately pleasing manner. A little taller than HHCH001, it is better in bigger rooms where scale is important.

HHCB001
Corner Bead

1900-1915
Victorian

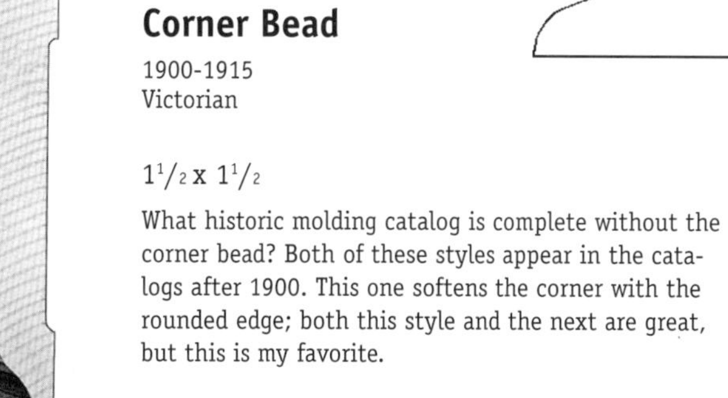

1$^1/_2$ x 1$^1/_2$

What historic molding catalog is complete without the corner bead? Both of these styles appear in the catalogs after 1900. This one softens the corner with the rounded edge; both this style and the next are great, but this is my favorite.

HHCB002
Corner Bead

1900-1915
Victorian

1$^1/_2$ x 1$^1/_2$

A popular Victorian mold, this corner detail doesn't appear in later catalogs. A great OG detail defines these edges.

HHCV001
Cove Molding

1870-1945

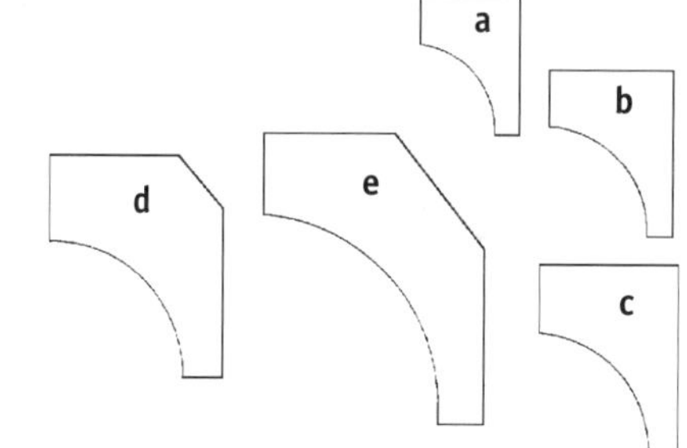

a $^5/_8$ x $^7/_8$
b $^3/_4$ x 1
c $^7/_8$ x 1$^1/_8$
d 1 x 1$^1/_2$
e 1$^1/_4$ x 2$^1/_4$

The cove mold, like the quarter round, had many uses and came in many sizes. Please note we offer five sizes below. Example b is the most popular size and is used under stools, or stair treads for a handsome finish.

HHEM001
Electrical Molding

1900-1915
Late Victorian

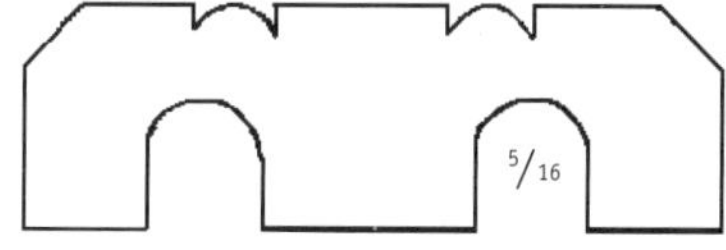

a $^1/_4$ x 1$^1/_8$
b $^9/_{16}$ x 1$^1/_2$

These moldings were available in catalogs from 1900-1915. I don't know if we will get a single order for one, but I wouldn't feel right presenting a historic molding catalog with out it.

HHEM002
Electrical Molding

1900-1915
Late Victorian

$^1/_2$ x 1$^3/_4$

This is a one-piece version of HHEM001.

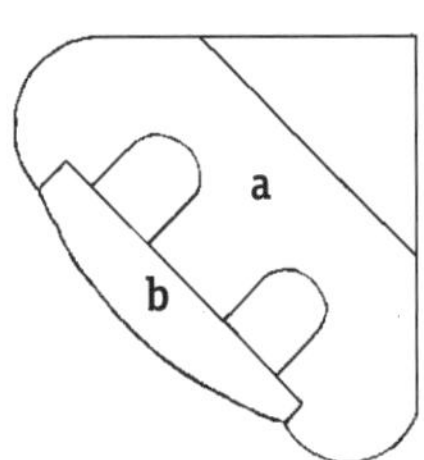

HHEM003
Electrical Molding
1900-1915
Late Victorian

a $^7/_8$ x $1^7/_8$

b $^1/_4$ x $1^1/_4$

When the electrical lines ran in the corner, they would use this corner piece. A great piece of history for your historic home, even if you never use it.

HHHR001
Half Round
1870-1945
Classical

$^1/_2$ — $2^1/_2$

in $^1/_2$ in. increments

The half round was available as a molding detail from 1870 into the 1940's. In each era it was used in different ways. In Victorian homes the half round was used to build up casings; later it was used as a decorative detail. It is available in many diameters.

HHLT001
Lattice Molding
1870-1945
Classical

$^5/_{16}$ x $1^1/_4$

The lattice mold was popular in catalogs, especially in the Victorian period. Many carpenters would order lattice stock and make their own custom lattice work. If you are using it for exterior purposes, let us know and we can run it in a more durable exterior grade wood.

HHLT002
Lattice Molding
1870-1945
Classical

$^1/_2$ x $1^3/_4$

A thicker and wider version of HHLT001 that would have been used by carpenters to frame out some latticework. It is a very versatile piece of wood that has limitless possibilities.

HHNM001
Nosing
1870-1945
Classical

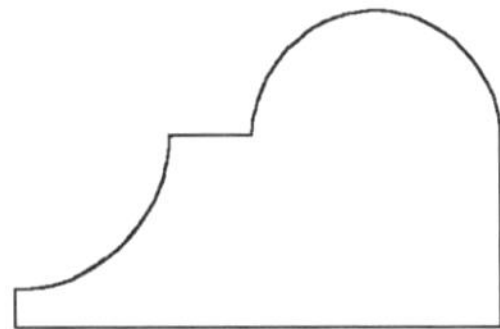

1 x $2^1/_4$

Nosings were a popular Victorian mold; this one was long lasting and appeared from 1870 through 1940. It had many uses—as a cap on the edge of stair treads, a countertop, or anywhere a flat board turns and goes down.

HHNM002
Nosing
1890-1915
Victorian

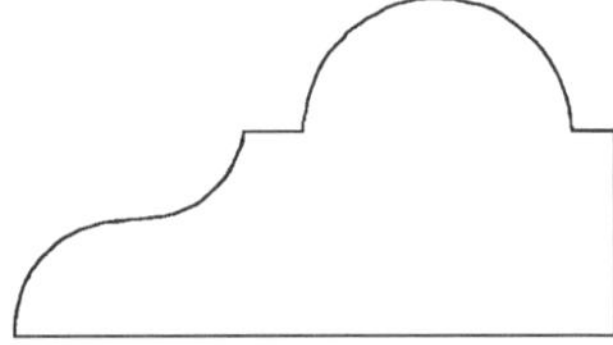

1 x $2^1/_2$

Nosing is similar to the HHNM001, but it uses the OG profile and has a small fillet on the back edge. This creates a nice shadow line that looks great painted or stained.

HHPW001
Pew Back Rail
1870-1915

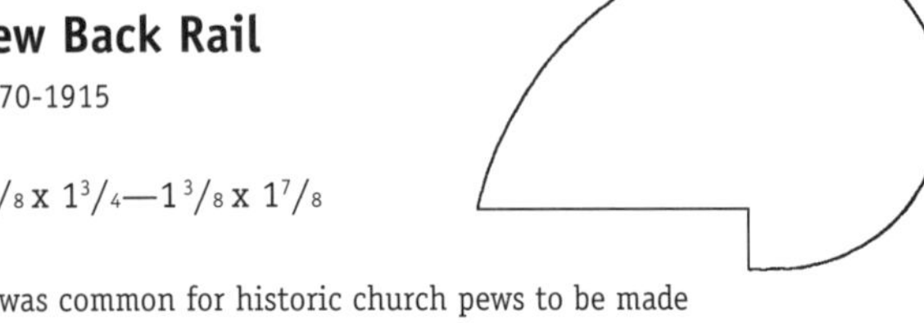

$1^1/_8$ x $1^3/_4$—$1^3/_8$ x $1^7/_8$

It was common for historic church pews to be made from large millwork companies, and the pew back cap was a standard mold through the Victorian period. Another molding that I bet you didn't expect to see when you picked up this catalog!

HHPS001
Partition Cap & Shoe
1890-1930

partition cap $1^1/_8$ x 2

partition shoe $^7/_8$ x $2^1/_2$

The partition cap and shoe is a period detail that I have a lot of experience with. Our headquarters are located in a 1921 warehouse, and we still have the original offices that were divided this way. If you are looking for an office space with a difference, allow us to bid your finish-out or remodeling. There are a lot of fun ways to configure the panels and great flexibility in how it is detailed out. These pieces are sold in 4x4 foot squares.

HHQR001
Quarter Round
1870-1945
Classical

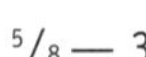

$^5/_8$ — 3
in $^1/_4$ in. increments after 1 in.

Another classic mold with a long life. Available from 1870 to 1940, the quarter round is used for many purposes. As mentioned in the Brick Molds section, large 2-3 inch quarter rounds were used as a brick mold. In Victorian times, the quarter round was used as base shoe mold in the $^3/_4$-inch and 1-inch sizes.

HHRB001
Return Bead
1870-1945
Classical

2 inches

The return bead was in historic millwork catalogs throughout this period. Like other classic molds, its uses changed over time. It was most popular in the Victorian age, when used as an inside and outside corner for wainscot.

HHSC001
Shelf Cleat
1900-1935
Turn of the Century

$^{13}/_{16}$ x $1^7/_8$

After I first saw these in the historic catalogs, I began searching for them in homes and found them everywhere. This was a popular detail used on built-ins and cabinets of historic homes up into the 30's. The rise of component cabinet construction caused this popular detail to be left out of catalogs.

HHSC002
Shelf Cleat
1900-1915
Late Victorian

$^{13}/_{16}$ x $2^1/_4$

Another cabinet detail for shelves, this piece was used with a cove or bracket underneath. Though simple, it is still a vast improvement over the slotted adjustable metal seen today.

HHSE001
Shelf Edge
1900-1915
Late Victorian

$^{13}/_{16}$ x $^7/_8$

This shelf edge is an historic way to dress up the edges of your shelves, and a nice way to finish kitchen cabinetry.

HHTH001
Threshold
1885-1925
Classical

$^3/_4$ x $4^1/_2$

The wood threshold appears in catalogs from 1890-1925. It stops appearing as manufacturers begin providing pre-hung doors and the local carpenter was not relied upon for this detail.

HHTB001
Transom Bar
1880-1900
Victorian

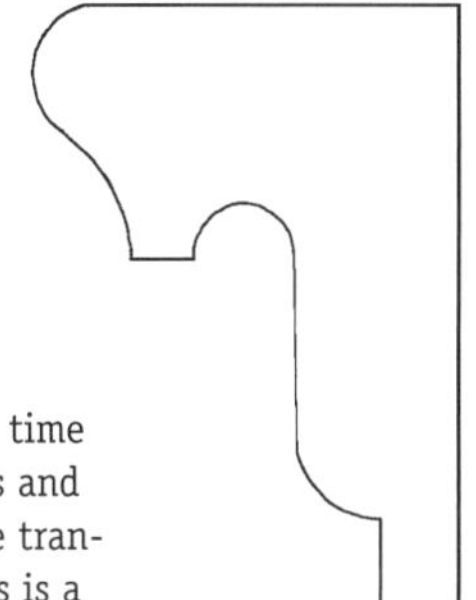

$1^3/_4$ x $2^1/_2$

The transom bar is a Victorian detail from the time when transom windows were used above doors and windows to increase ventilation. I also see the transom detail used in new homes effectively. This is a great way to add authenticity to any project.

HHTB002
Transom Bar
1880-1900
victorian

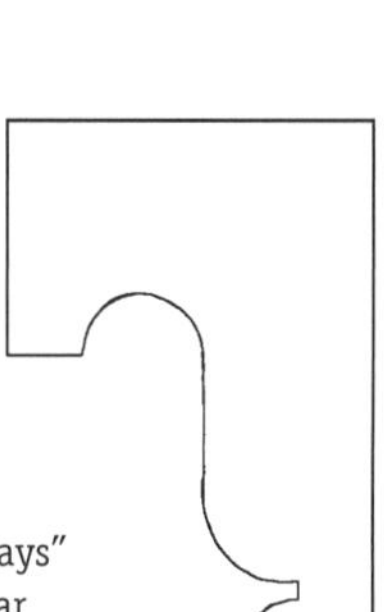

$1^3/_4$ x $2^1/_2$

These transom bars remind us of the "good old days" before air conditioning cooled our homes. This bar has an attractive bead detail at the bottom and is a nice period-appropriate look.

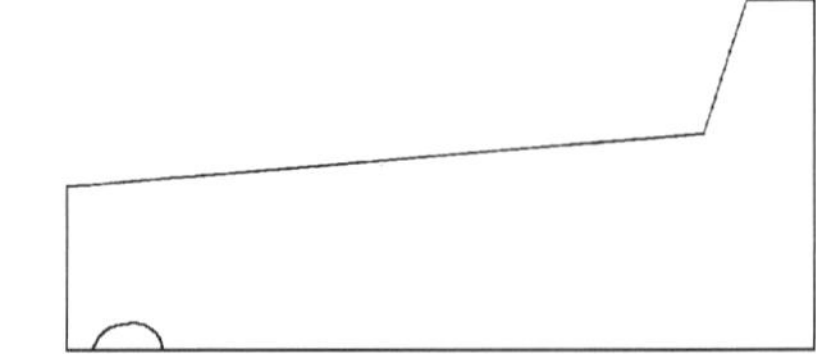

HHTB003
Transom Bar

1880-1900
Victorian

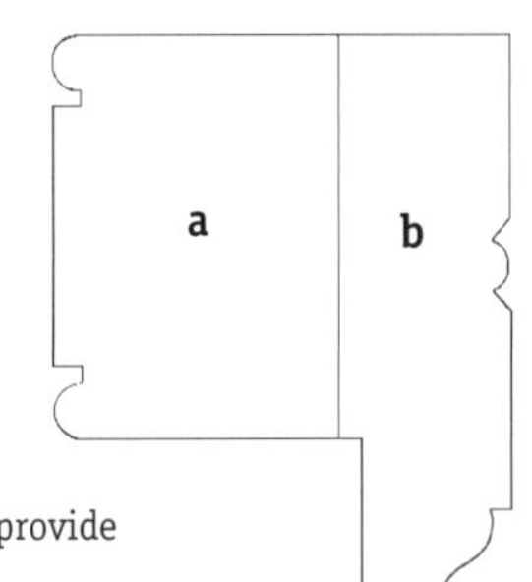

a $1^3/_8$ x $1^7/_8$

b $^{13}/_{16}$ x $2^3/_4$

A two-piece transom bar that laps over to provide a nice look with appealing detailing.

HHWT001
Water Table
or Drip Cap

1900-1925

$1^3/_8$ x 2 or 3 or 4 or 5 inches

There is no better detail for a wood-sided house than the water table or drip cap at the base. It is a form-pleasing piece that visually grounds the house and performs the important function of keeping water from getting into it.

Molding Packages

MOLDING PACKAGES

MOLDING PACKAGE HHMP007

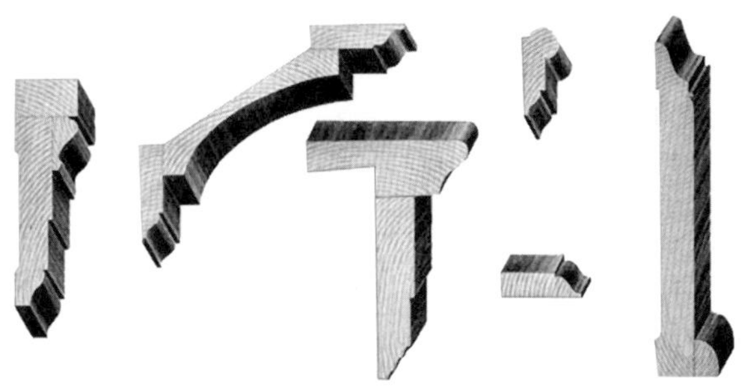

HULL
HISTORICAL
MOLDING

Molding Packages

We offer 14 molding packages that represent four great periods of architectural history: Victorian, Turn of the Century, Arts & Crafts, and Period Revival. In addition to the wonderful designs in these pages, you will also notice that the packages literally look different; this is because we are showing them exactly how they appeared in the historic catalogs.

The Victorian package is probably the most customized. In showing them as they were presented in the original molding catalogs, please realize there is custom work involved to make this fit in your home. We believe there is simply no other way to capture the ornate and wonderful details that make these molding packages special. Just as the original molding companies could do any work, so can we and we look forward to completing your molding package with these and other appropriate details.

The-turn-of-the-century packages are a mix of the Victorian and Arts & Crafts styles. They blend a little ornamentation with simple styling for a great look.

The Arts & Crafts packages are typical of what would be found in Bungalows, Craftsmen and Prairie style houses. This classic simplicity is wonderful and unique.

The Period Revival packages would be found in the Colonial Revival as well as the Old World revival style houses like Mediterranean, Spanish, Tudor, French and Mission styles. They work in historic homes as well as new homes.

Some packages are shown without a base or without a crown. If any moldings are not shown that you would like, we can recommend a complimentary molding for you to complete the package. We can also provide a complete printout of the package so that you can label each piece and tell us the footage you need for every part.

HHPM001
Molding Package

1890
Victorian

a Plinth
b Base
c Shoe
d Base Cap
e Header Block
f Casing
g Stool
h Custom Header

A great molding package that really expresses the ornate joy of Victorian architecture. Note: this is one of the custom packages which has items not found in the standard catalogs. To complete this room requires custom ordering from your plans.

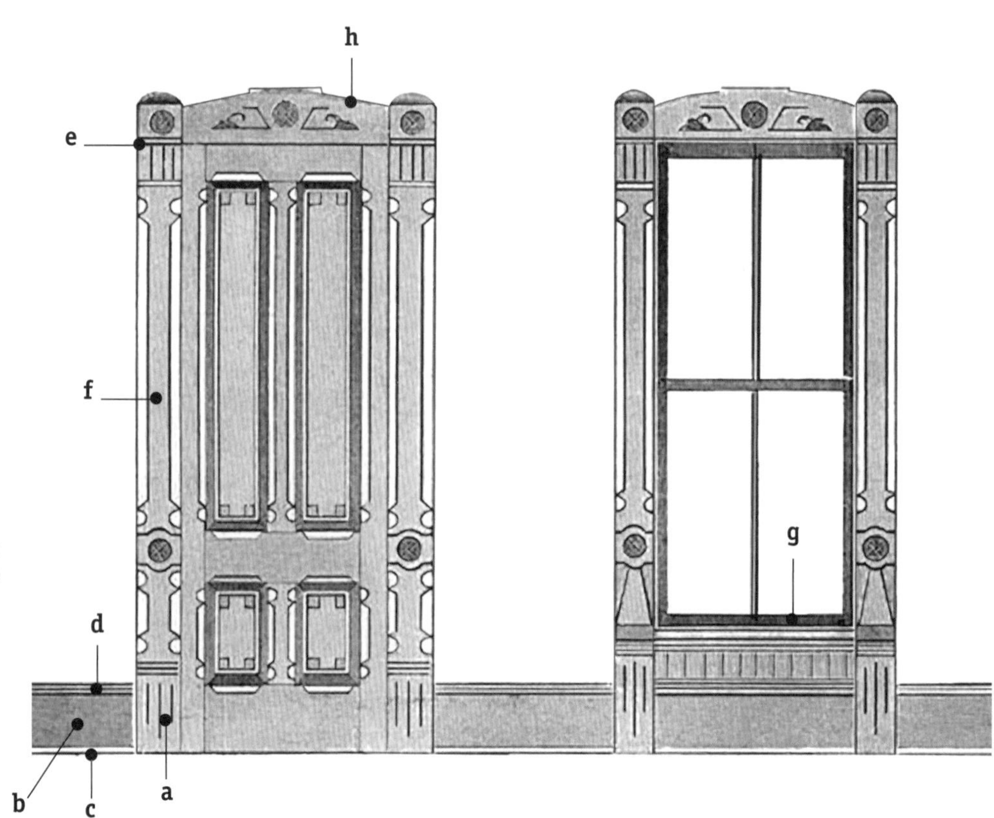

HHPM002
Molding Package

1890
Victorian

a Plinth
b Base
c Base Cap
d Shoe
e Rosette
f Header Block
g Casing
h Window Stool
i Wainscot Cap
j Wainscot Boards

A Victorian package "by the book." Everything here comes from the catalogs and is a great example of what standard moldings can become in Victorian architecture. The wainscoting is a custom detail that can be added to your room if you desire.

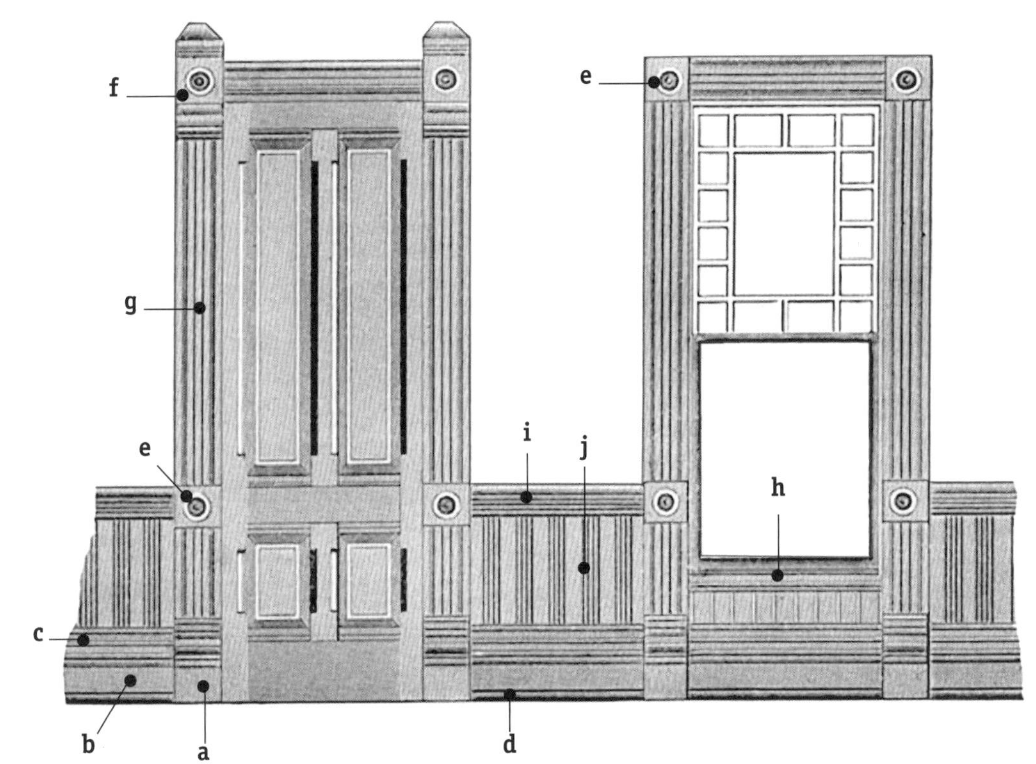

HHPM003
Molding Package

1890
Victorian

a Plinth
b Base
c Shoe
d Base Cap
e Header Block
f Casing
g Stool
h Custom Header
i Custom Header
j Header Block

A combination of custom and standard products, this package would make a handsome statement in any home. Our custom shop can make the panel under the window as well as make the door if you would like to complete the look.

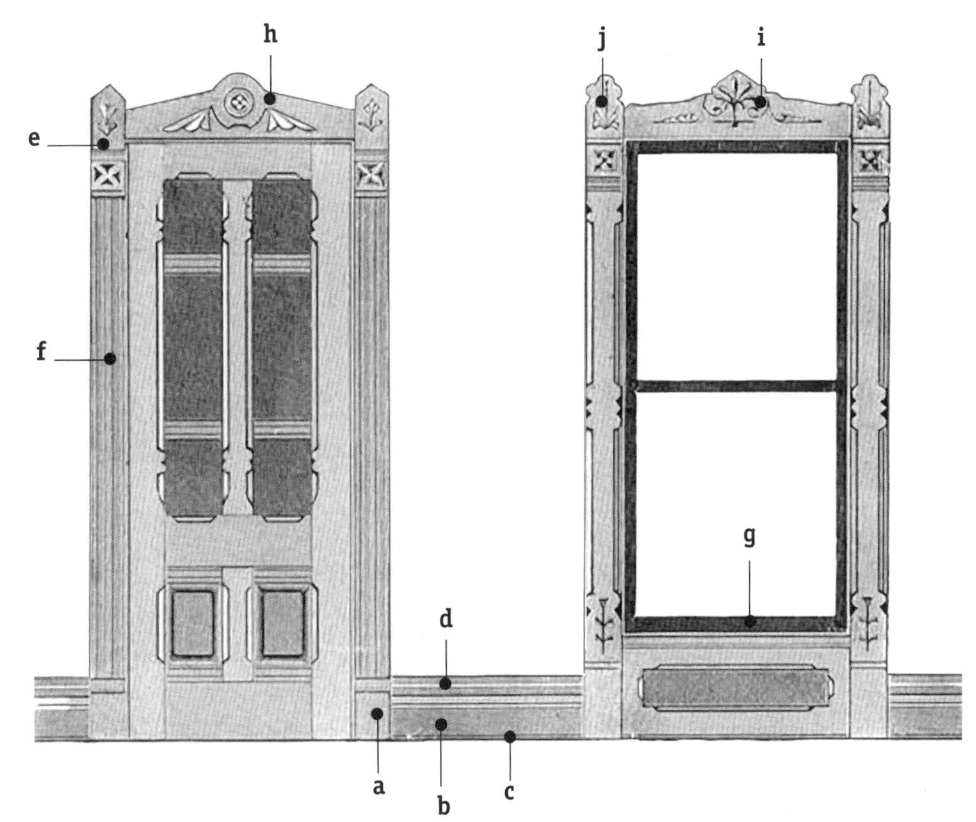

HHMP004
Molding Package

1890
Victorian

a   Plinth
b   Base
c   Shoe
d   Base Cap
e   Header Block
f   Casing
g   Stool
h   Custom Header
i   Custom Header
j   Custom Casing
k   Custom Plinth
l   Casing
m   Custom Ornament

One of the hallmarks of the Victorian millwork houses was their ability to do custom work. In that vein, we offer another custom package filled with ornamentation that we would love to quote for your project.

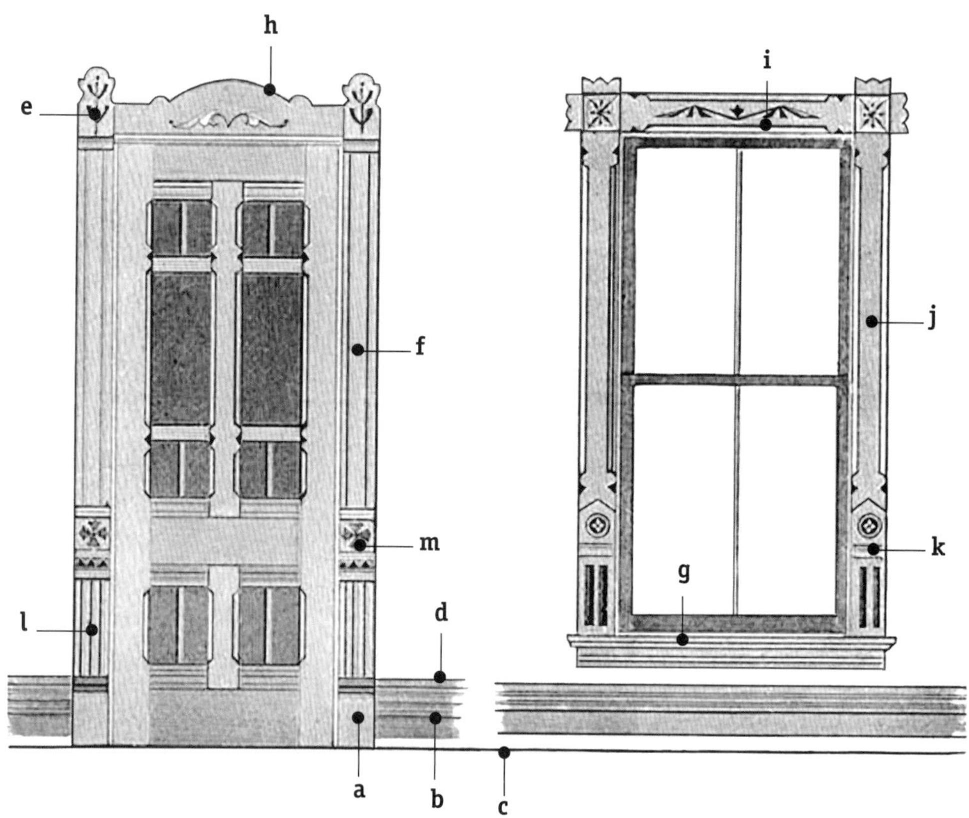

HHMP005
Molding Package

1903
Turn Of The Century

a   Plinth
b   Base
c   Shoe
d   Casing
e   Header Stop
f   Header Cap
g   Custom Bead & Reed
h   Window Stool

This Turn of the Century package from a 1903 catalog represents the changing styles of Arts & Crafts and Victorian. The cap on the doors and windows has a decorative bead and reed detail that really works well with the plain casing and plinth.

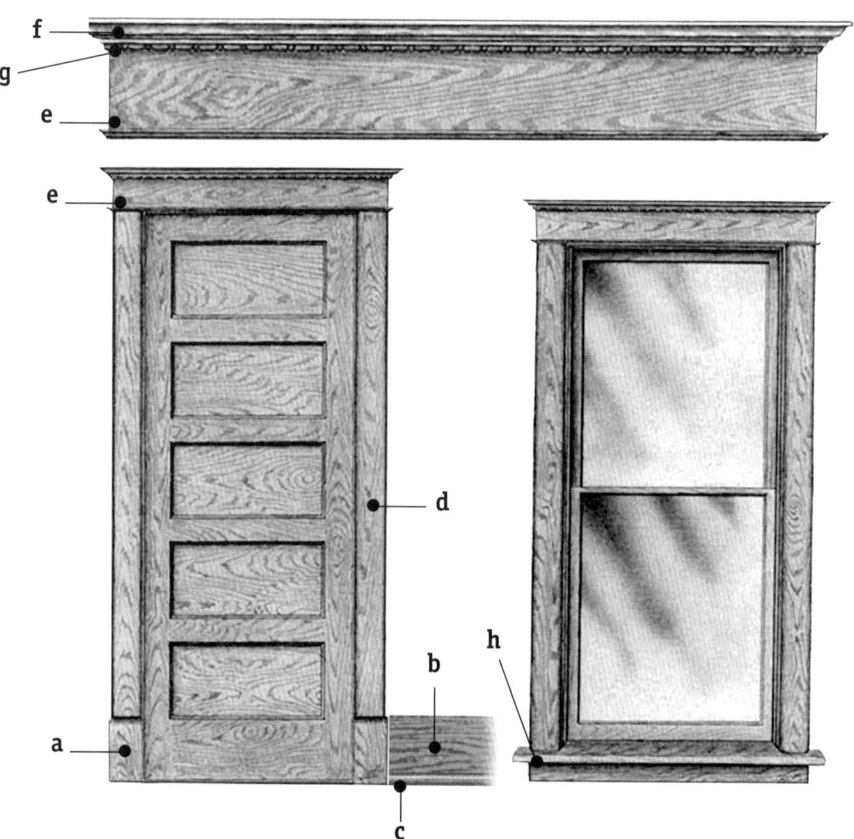

HHMP006
Molding Package

1903
Turn Of The Century

a   Plinth

b   Base

c   Shoe

d   Base Cap

e   Casing

f   Header Stop

g   Header Cap

h   Stool

The egg-and-dart detail in the cap really tops off this package. The embossed wreaths we offer as an option are very similar but not exactly like those shown.

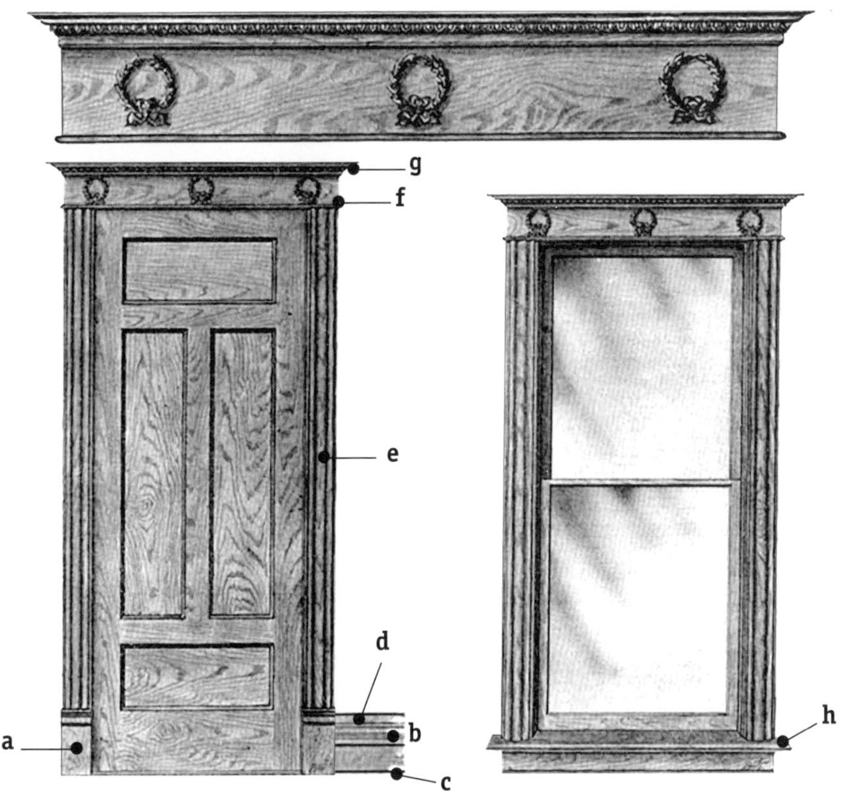

HHMP007
Molding Package

1903
Turn Of The Century

a   Plinth

b   Base

c   Shoe

d   Casing

e   Stop

f   Header Cap

g   Window Stool

Many Turn of the Century packages had a dental header for windows and doors. This great detail works well in today's homes and can really add style to any Colonial Revival home.

HHPM008
Molding Package

1900-1920
Arts & Crafts

a Plinth
b Base
c Shoe
d Casing
e Window Stool
f Optional Wainscot
g Optional Wainscot Cap
h Picture Mold

This Arts & Crafts packages combines classic Arts & Crafts detailing like the header that overhangs the casings and the tall wainscoting (optional). The plinth is plain and a reminder of classical detailing. It is a simple but strong package.

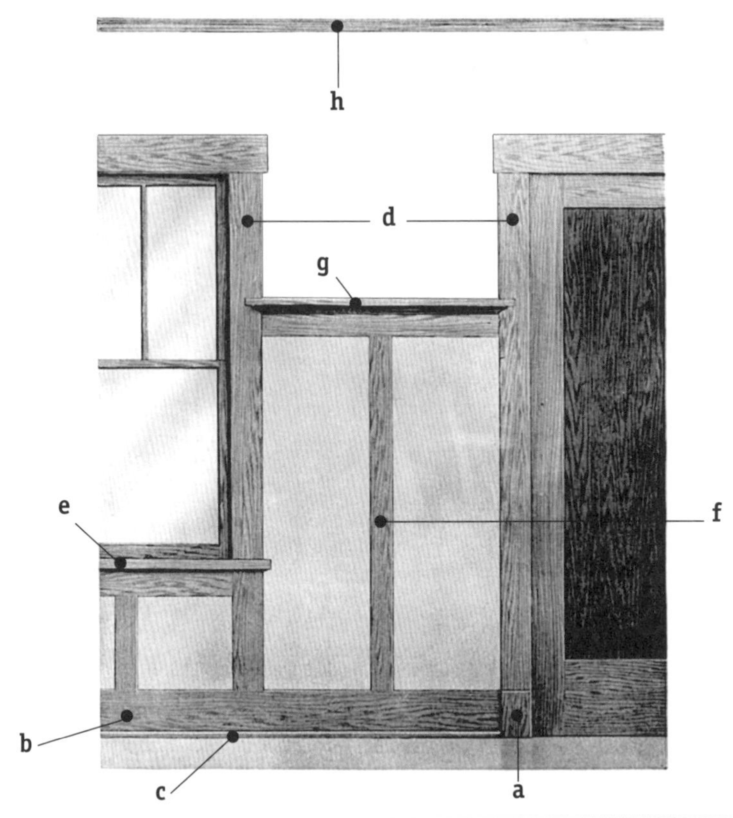

HHPM009
Molding Package

1900-1920
Arts & Crafts

a Casing
b Back Band
c Base
d Base Shoe
e Wainscot Cap
f Picture Mold
g Window Stool

This is another package which works great with a simple décor theme like Arts & Crafts. There is a back band on the outside edge of the casing which provides some boldness.

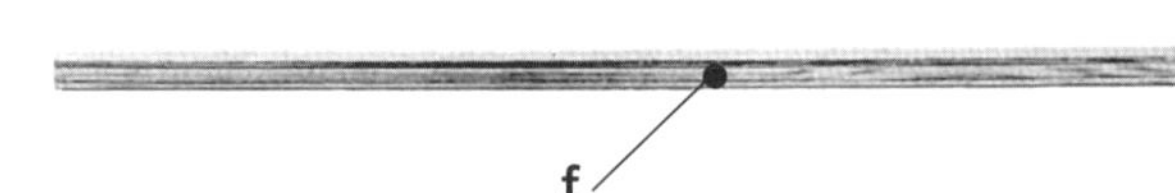

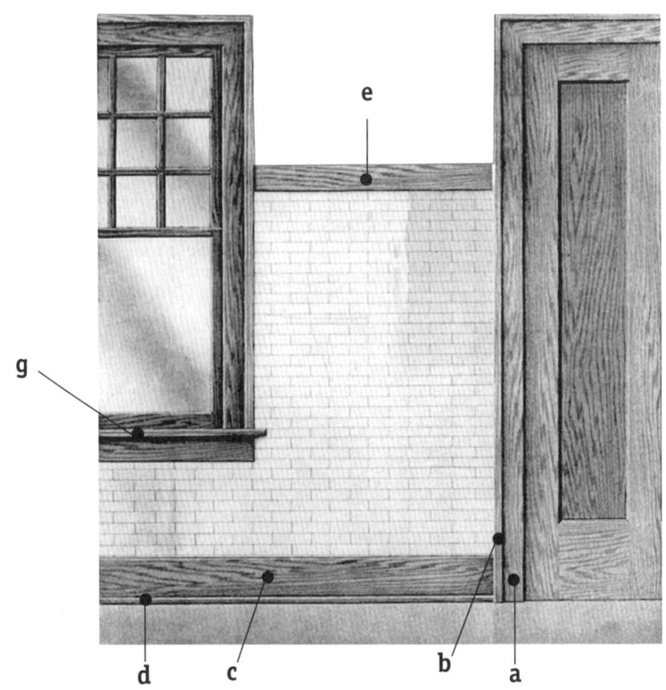

HHPM010

Molding Package

1900-1920
Arts & Crafts

a Plinth

b Base

c Shoe

d Casing

e Stool

f Apron

g Picture Mold

Another Arts & Crafts package which includes a popular casing. As on most Arts & Crafts packages, there is no crown—only a picture mold which was common on historic bungalows and Craftsman homes.

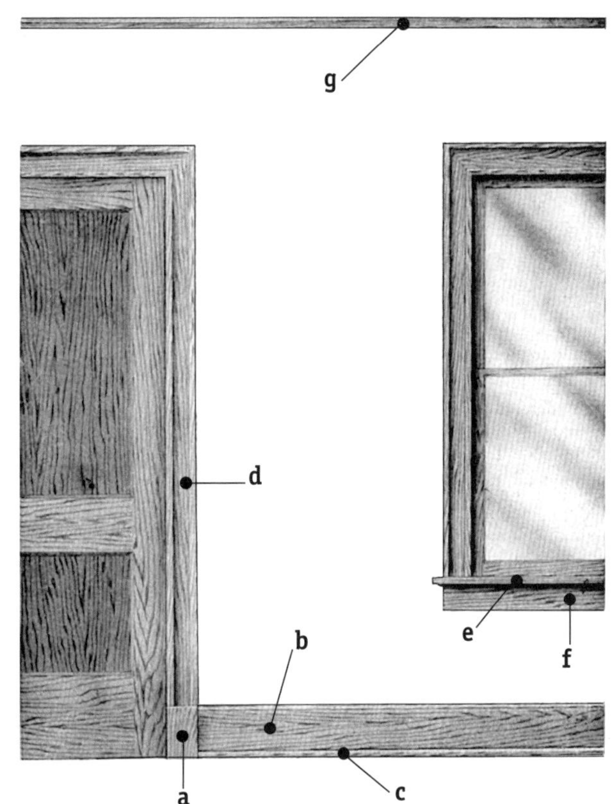

HHPM011

Molding Package

1920's-1930's
Period Revival

a Base
b Shoe
c Stop
d Casing
e Stool
f Apron
g Picture Mold
h Crown (2-part)

This is a Colonial Revival package with real design merit. An appropriate package for any Revival style home.

HHPM012
Molding Package

1920's-1930's
Period Revival

a Casing
b Back Band
c Stool
d Apron
e Crown
f Panel Mold
g Stop
h Picture Mold
i Base
j Shoe

Another wonderful Period Revival molding package that works especially well on the Old World style home where walls were paneled out as shown here. The crown is attractive and would work on either a new or historic home.

HHPM013
Molding Package
1920's-1930's
Period Revival

a Casing
b Back Band
c Header Cap
d Stop
e Crown
f Panel Mold
g Picture Mold
h Apron & Cove
i Stool
j Base
k Shoe
l Base Cap

A Classical Revival package, with a header cap that really finishes off the doors and windows. Note the panel mold that adds interest to the crown. Combined with the base cap, it provides the package with real finish and flair.

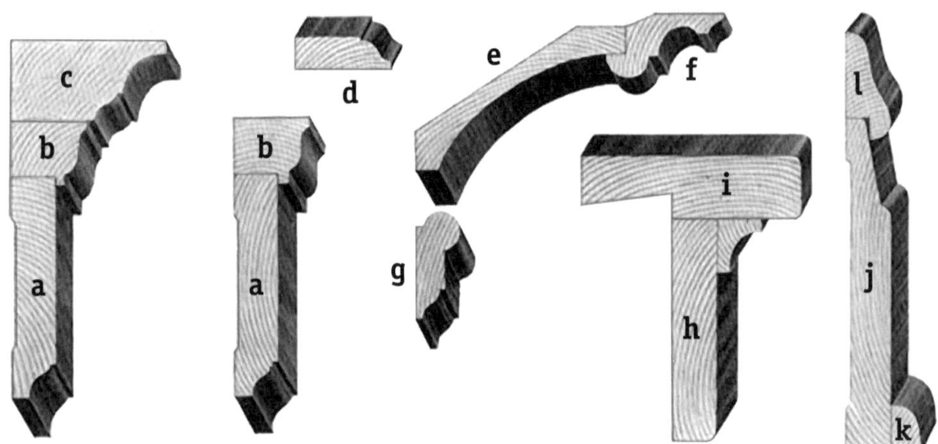

HHPM014

Molding Package

1920s-1930s
Period Revival

a   Casing
b   Header Cap
c   Stop
d   Apron
e   Stool
f   Crown
g   Picture Mold
h   Casing
i   Base & Cap
j   Shoe
k   Plinth

Another attractive Period Revival package. The wainscot with panel mold design is a great finishing detail either painted or stained. The plinth block is a holdover from previous periods and gives this package extra charm.